As if by Chance

Sketches of Disruptive Continuity in the Age of Print
from Johannes Gutenberg to Steve Jobs

Kevin Reed Donley

Copyright © 2023 Multimediaman
Southfield, Michigan
www.multimediaman.net

Fulton Books
Meadville, PA

Published by Fulton Books 2023

ISBN 979-8-88505-792-9 (paperback)
ISBN 979-8-88982-152-6 (hardcover)
ISBN 979-8-88505-793-6 (digital)

Printed in the United States of America

To the men and women—pressroom and bindery workers, prepress technicians, maintenance people, production managers, customer service staff, estimators, office staff, sales representatives, and business owners—that I have worked with over the past forty years in the graphic arts, commercial printing, and publishing industries. I include here those who have become lifelong friends as well as those who I did not have the opportunity to get to know personally. All of you have inspired me with your knowledge, skill, and commitment to excellence, and I could not have completed this project without you.

Contents

Preface

This project began in June 2008 when I traveled from Detroit to Düsseldorf, Germany, and attended DRUPA, the international printing and paper exposition. During that trip, I witnessed significant change in printing technology, including a demonstration of the first ink-jet web press as well as the impact of many digital processes on the industry. While I was in Germany, I took a train ride to Mainz and spent an afternoon at the Gutenberg Museum. Among the artifacts I saw there were copies of the Bible printed by Johannes Gutenberg in 1455 and a live demonstration of the fifteenth century inventor's handheld mold for metal-type casting.

At that time, my professional and personal life were being altered by the smartphone, wireless broadband, cloud services, video streaming, and social media platforms such that I decided to record my seven days in Germany in a blog that I called Multimediaman. To me, this word not only defined my identity as someone living and working with my back foot in the old analog printing world and my front foot in the new digital age of the future; it also described what we are all becoming. By embracing electronic publishing in its various forms, billions of people around the globe are transforming human culture into a vast online multimedia repository. It occurred to me that this digital likeness of man will one day be reviewed by historians as the start of a great modern transformation much in the same way they now study the incunabula for an understanding of the revolutionary changes in society associated with the Renaissance and the birth of print.

In 2009, I began a review of the evolution of the five and a half centuries of printing by writing sketches of the lives and inventions of some of its greatest technical innovators. In between these biographical studies, I also wrote about our multimedia world—from mobile technologies and the creation of e-paper to the genesis of the graphical user interface and the emergence of digital photography and big data—from the perspective of an early adopter of these technologies. These alternating studies, published online over the next eight years at Multimediaman.net and in the pages of *Graphic News*, the printed monthly magazine of the Printing Industries of Michigan, are the subjects of both the chapters and appendices of this volume.

While I could have stopped there, I was perplexed by the persistence of rhythmic parallels in the technical, socioeconomic, and cultural developments in and around previous generations of printing. One of these trends was a pronounced repetition of the attribution of "accidental invention" to major innovations in the history of print technology. I was curious about how this phenomenon was experienced by inventors more than once during critical points of technological transition, and I sought to more fully understand the role of happenstance in the evolution of printing history, if not in technical progress generally.

Another example of a resonating cycle was the accelerating trend toward greater distribution of information and knowledge among the public that began with Gutenberg's printing revolution. The loss of control over printed forms of human culture such as books, broadsheets, pamphlets, and newspapers by the ruling powers of the medieval church and the autocratic state, beginning in the fifteenth century, contributed to a vast expansion of literacy and a drive by the reading public for democratic forms of government that more fully corresponded to the changes in the social environment brought on by capitalism, manufacturing, and industrialization.

With the emergence of electronic media associated with the desktop computer and Steve Jobs, however, this quantitative accumulation of mass information distribution has taken on a qualitatively new and profoundly transformative expression as every consumer of information and knowledge is now also a publisher. With this transformation, the present-day boundaries of monopoly control over publishing, news and the circulation of information by media corporations and the national state have been breached, and the foundation for a new era of a global socioeconomic and democratic political identity has been laid. As opposed to the centuries of comparatively slow development of printing technology, which planted seeds around the world that grew, penetrating the whole of human culture and brought to bear its latent ideological, social, and political force during the Enlightenment and national democratic revolutions of Europe and America in the late 1700s, it appeared to me that the ultimate societal impact of our new digital media age could emerge in a much shorter time frame, perhaps decades, but this transformative potential existed today in embryonic form and had clearly not yet been realized.

These topics are examined in the introduction, where I define and expand upon the theory of technological evolution called disruptive continuity, and they are also the basis of the title of this book.

*****

The term *technology* and its derivatives are used repeatedly in this book. In some instances, words such as *apparatus, artifact, equipment, invention, machine,* and *tool* have been used interchangeably with *technology.* As employed here, technology encompasses the overall and general category of human productive forces and culture as well as the material object of invention and the ideas and social processes by which it is created and used. While it is a very flexible term, *technology* is also decidedly a modern one. Until well into the industrial revolution, the concept of *technology*—in the sense that it has been used beginning in the late twentieth century—did not exist. Among the difficulties in writing about the history of technical progress, especially a review that spans centuries and encompasses several socioeconomic eras, is the danger that the modern usage of a word, such as *technology,* will be falsely projected onto the ideas of the past when there was not and could not have been such a concept.

In his excellent book *Technology: Critical History of a Concept*, Eric Schatzberg explains that the *-ology* suffix of the word suggests it should refer to an academic field of study, as in "Massachusetts Institute of Technology," for example. However, in its present-day usage, *technology* "refers more to things than ideas, to material practices rather than a scholarly discipline." Schatzberg writes:

> before the 1960s, the term *technology* was largely confined to scholarly discourse. It first appeared in sixteenth-century academic Latin as *technologia*; by the seventeenth century, the term was listed in at least one dictionary in its English form. For the next few centuries, *technology* occupied recondite corners of scholarship, occasionally bubbling to the surface in some narrow context only to disappear again, until it found a place in the consciousness of Anglo-American scholars after World War I. Even then, *technology* remained uncommon among nonspecialists, except in reference to technical education. Not until the 1960s did *technology* rise to the status of keyword in popular discourse.

As Schatzberg's historical exposition shows, the new forms of machinery associated with the twentieth-century electronics revolution, combined with the public adoption of sophisticated hardware and software systems—personal computers, networks, digital communications, broadband wireless services, mobile devices, and social media publishing platforms—are the industrial backdrop to the transformation of the term *technology* from a concept only found in narrow academic circles to a word used by billions of people to describe generically the tools they use on a daily basis. The term embodies the mass form of the instruments themselves, driven by the socioeconomic changes present in the transition from print to electronic media culture.

Introduction

*Everything existing in the universe is
the fruit of chance and necessity.*
　　　　　　—Democritus, circa 400 BC

*Necessity is blind only so long
as it is not understood.*
　　　　　　—G. W. F. Hegel, 1817

It is notable that some inventions in the history of print technology are recorded as having been achieved by chance. In accounts written at the time of the inventions as well as in historical studies, major breakthroughs in printing have been attributed to accidental events. Much in the same way schoolchildren are taught that the natural scientist Isaac Newton discovered the law of gravitation after an apple fell from a tree upon his head, significant inventions in the history of printing are said to be the result of lucky mistakes.

Perhaps the two most well-known examples of this phenomenon are found in accounts of the late-eighteenth-century invention of lithography by Alois Senefelder and the early-twentieth-century invention of offset printing by Ira Washington Rubel. In both cases, the technical advances made by the inventors are often explained as having been accidental.

Here are two citations:

> Lithography was invented around 1796 in Germany by an otherwise unknown Bavarian playwright, Alois Senefelder, who accidentally discovered that he could duplicate his scripts by writing them in greasy crayon on slabs of limestone and then printing them with rolled-on ink (Department of Drawings and Prints, The Metropolitan Museum of Art, October 2004).

> Offset printing, also called offset lithography, or litho-offset, in commercial printing, widely used printing technique in which the inked image on a printing plate is printed on a rubber cylinder and then transferred (i.e., offset) to paper or other material. The rubber cylinder gives great flexibility, permitting printing on wood, cloth, metal, leather, and rough paper. An American printer, Ira W. Rubel, of Nutley, New Jersey, accidentally discovered the process in 1904 and soon built a press to exploit it (the editors of *Encyclopædia Britannica*, July 1998).

Readers of these passages would not be blamed for thinking that Senefelder of Bavaria, Germany, in 1796 and Rubel of Nutley, New Jersey, in 1904 were the beneficiaries of pure luck or that they fortuitously stumbled their way into print technology history. However, this would be an incorrect or, at best, an incomplete way of understanding the contributions of these two innovators.

Why does the word "accidentally" appear in the above accounts of historic inventions that took place more than one hundred years apart and, together, established what is known as offset lithography, a technology that revolutionized the printing industry and remains today the dominant method of transferring ink to paper? Why is it that stories of accidental invention—even from authoritative sources like the Metropolitan Museum of Art and *Encyclopædia Britannica*—persist for both men, despite ample evidence that Senefelder and Rubel were in pursuit of innovation and striving to improve the printing process through the methods of ingenuity, experimentation, and science that prevailed during their respective lifetimes?

It will take an investigative journey to find the answers to these questions. While it may be a fact of popular interest that Senefelder and Rubel are known as much or more for the accidental way they arrived at their achievements than they are for the achievements themselves, it is also a fact that invention by happenstance has occurred in history more often than is generally known. Since the "accidental" attribution tends to overshadow and mystify the progress attained, in printing as well as other fields, it is instructive to examine these two inventions in their socioeconomic context and to locate the place of Senefelder and Rubel within the whole history of printing. This examination shows that their accomplishments were decisive steps in the transformation of printing from a handcraft to a mass-manufacturing industry in the nineteenth and twentieth centuries.

To untangle the riddle of accidental invention in the cases of Senefelder and Rubel, it is necessary to (1) investigate the historical record and review the facts of what is known about the men and how they invented lithography and offset printing; (2) look outside print technology and into the prevalence of "serendipity" more broadly in the history of scientific and technological discovery; (3) explore the source of the need for legends of accidental discovery in human progress; (4) make a theoretical analysis of the two-sided and contradictory content of "accidents" in general; and (5) return to Senefelder and Rubel and show how their inventions were manifestations of disruptive continuity in the history of printing.

The theory of disruptive continuity is a conceptual framework for understanding the historical development of printing technology—as one of the most important elements of modern technical progress—because it acknowledges that each stage of innovation owes much to the accomplishments of others that came beforehand; that breakthrough inventions could not have taken place without innumerable connections to the past. At the same time, disruptive continuity holds that each new achievement represents a sharp departure from the past, that it is a transition point forward that expresses the future in ways that were previously impossible and could not have been accomplished but for the spark embedded in the new generation of technology.

As this introduction will go on to explain, it is at the nexus point of discontinuity from the prior gradual progression and the moment of a leap into the future that the phenomenon of accidental invention occurs. To understand how unanticipated events, which are rooted in antecedent accomplishments, can and do become transformed into significant innovations is to understand the dialectical process by which the old technology is superseded by that of an entirely new era of progress.

Finally, by developing the historical-technical analysis of nearly six centuries of print communications contained in this book, significant conclusions can be drawn about the process by which the age of ink-on-paper media associated with Johannes Gutenberg is being displaced by the new age of electronic media associated with the breakthrough technology of desktop publishing identified with Steve Jobs. Based on the theory of disruptive continuity, it will be shown that the future point at which the new digital environment of online, mobile, social media, and virtual reality systems will supersede the mechanical, and analog world of printing is dependent upon much more than the evolution of the technology itself.

*****

This investigative journey begins with an examination of the work of our two print technology innovators who are frequently recognized as accidental inventors. It is fortunate that, in the case of Senefelder, an account written by the inventor himself is available and, in the case of Rubel, there exist two technical explanations, an anecdotal account and a posthumous tribute to the inventor written shortly after his death by a close business partner.

The invention of lithography

In 1817, at the urging of his colleagues, Alois Senefelder wrote down the story of his life along with a detailed description of how he invented lithography by experimental methods. He also provided a step-by-step technical guide for those wishing to learn and practice the art also known as "printing from a stone" or "stone printing." Senefelder's account was published one year later in the German volume entitled *Vollständiges Lehrbuch der Steindruckerey* (A Complete Course of Lithography). The work was translated into English by J. W. Muller and published by the Fuchs & Lang Manufacturing Company in New York in 1911 as *The Invention of Lithography*.

The relevant passages from the 1911 English text are found in the first chapter, "Section I: History of Stone Printing, Part I: From 1796 to 1800." As mentioned in the above quote from the Metropolitan Museum of Art, the young Alois Senefelder was an aspiring playwright and was motivated to start a printing firm so he could publish his own works. Senefelder wrote that he was familiar with the procedures of the letterpress printing process of his day, "I had spent many a day in the establishments," and that "it would not be hard for me to learn." Senefelder also had a "desire to own a small printing establishment myself" because—having studied both public finance and law for three years at the University of Ingolstadt—he wanted to "earn a decent living" and "become an independent man" by going into business.

However, it was economic reality—a lack of the financial resources required to become a printer—that drove Senefelder down the path of innovation. As he wrote, "If I had possessed the necessary money, I would have bought types, a press, and paper, and printing on stone probably would not have been invented so soon. The lack of funds, however, forced me to other expedients."

Senefelder gave details of three different approaches he took to replicate the letterpress method without the ability to purchase the technologies readily available to others with the requisite capital resources.

These were the following:

1. To etch letters in steel and then "impressing them on pear wood, in which the letters would show in relief, somewhat like the cast type of the book printers, and they could have been printed like a woodcut." He abandoned the approach, "I had to give up the whole thing through lack of implements and sufficient skill in engraving."

2. To purchase "enough types to set one column or folio" and transfer the letters "to a board covered with soft sealing wax and reproduce the relief plate thus obtained in stereotype form." Although this method was a technical success—especially after he began "mixing finely powdered gypsum with the sealing wax" and "made the latter harder than the ordinary type composition"—Senefelder was unable to move forward because "even this exceeded my financial power." He gave up on this plan, "especially as I had conceived a new one during my experiments."

3. To learn "to write out ordinary type letters exactly but reversed" with "an elastic steel pen on a copper plate covered in ordinary manner with etching surface," and these plates would be given to copper-plate printers for the press work. Here, Senefelder had difficulties because, though he quickly learned the skill of writing in reverse, "I could not correct the errors made during writing" because the "accessories of copper-plate engravers, especially the so-called cover varnish, were quite unknown to me."

Senefelder then "labored desperately to overcome the difficulty" and tried three sub-methods within this "elastic steel pen" approach:

1. Having "attained much chemical knowledge" during his days as a student, Senefelder began working with "spirits of wine and various resinous forms" and "oil of turpentine and wax" as methods for making corrections on the copper plate. However, he abandoned these materials because the chemi-

cal solution frequently became heavily diluted and "caused it to flow too much and dissolve the etching surface, at which time several well-done parts of the engraving were ruined."

2. Still determined to work with copper plate, Senefelder experimented with a wax and soap mixture as a material that could be used for correcting mistakes. He used "a mixture of three parts of wax with one part of common tallow soap, melted over the fire, mixed with some fine lampblack, and then dissolved in rainwater, gave me a sort of black ink with which I could correct faulty spots most easily." But this path "presented a new difficulty" in that he had only a "single little copper plate," and after he "pulled proofs at the house of a friend who possessed a copper-plate press," he had to spend "hours again laboriously grinding and polishing the plate, a process which also wore away the copper fast."

3. To get around the limited copper-plate resources, Senefelder transitioned to experimentation with "an old zinc plate of my mother's" that was "easier to scrape and polish." However, "the results were very unsatisfactory" because the "zinc probably was mixed with lead," and he did not have a "more powerful acid" that could penetrate it.

Finally, Senefelder moved on to transferring a printed image to paper based on "a handsome piece of Kelheimer stone." He explained, "The experiments succeeded, and though I had not thought originally that the stone itself might be used for printing (the samples I had seen hitherto of this Kelheim limestone were too thin to withstand the pressure exerted in printing), I soon began to believe that it was possible. It was much easier to do good work on the stone than on the copper."

He began working "in order to use the stone just like copper" and trying "all possible kinds of polishing and grinding without attaining my purpose completely." Senefelder wrote that moving from copper or zinc plate to printing from a limestone did not immediately result in the invention of lithography, "I had invented little that was new but simply had applied the copper-plate etching method to stone."

And furthermore, "I was not the first discoverer of stone etching nor of stone printing, and only after I made this new discovery which I will describe now, which led me from the engraved to the relief process, with my new ink might I call myself the inventor of an art."

Amid his detailed survey, Senefelder made it clear he decided to write his account in 1817 to set the record straight, "I have told all of these things fully in order to prove to the reader that I did not invent stone printing through lucky accident but that I arrived at it by a way pointed out by industrious thought."

However, he went on to say his experiments with etched, that is, mechanical and relief but not yet chemical, processes on stone "were entirely checked by a new, *accidental* discovery. Until now, I had invented little that was new but simply had applied the copper-plate etching method to stone. But this new discovery founded an entirely new form of printing, which basically became the foundation of all succeeding methods." (Emphasis added)

Senefelder then recounted his well-known story of the accidental invention:

> I had just ground a stone plate smooth in order to treat it with etching fluid and to pursue on it my practice in reverse writing when my mother asked me to write a laundry list for her. The laundress was waiting, but we could find no paper. My own supply had been used up by pulling proofs. Even the writing ink was dried up. Without bothering to look for writing materials, I wrote the list hastily on the clean stone, with my prepared stone ink of wax, soap, and lampblack, intending to copy it as soon as paper was supplied.
>
> As I was preparing afterward to wash the writing from the stone, I became curious to see what would happen with writing made thus of prepared ink…
>
> My further experiments with this relief plate succeeded far better than my previous ones with

etched letters. The inking in was much easier, and hardly one quarter of the force was necessary for making impressions. Thus, the stones were not so liable to crack, and what was the most important for me, this method of printing was entirely new, and I might hope to obtain a franchise and even financial aid.

It would take further experimentation with the stone by Senefelder to finally arrive at the invention of lithography, "Even this method, new in 1796, still was purely mechanical in its purpose, whereas the present printing method, which I began in 1799, may be called purely chemical."

The following can be drawn from the above summary of Senefelder's own account of his invention: (1) Senefelder began in 1796 by experimenting and practicing with multiple materials and chemicals as he sought to develop an affordable mechanical printing process that was less-capital intensive than the letterpress method; (2) he insisted he did not invent lithography "through a lucky accident" but by way of "industrious thought"; (3) he stated his efforts to come up with an alternative mechanical method to letterpress "were entirely checked by a new, accidental discovery"; (4) he told the story of how, while working with a limestone as a mechanical image transfer base, he wrote a laundry list upon the stone and from here new possibilities then occurred to him; and (5) it would take three more years of further experimentation with the limestone before the "purely chemical" printing method was discovered in 1799 that become known as lithography.

It is highly significant that in his own account, Senefelder presented two different and internally contradictory explanations for how he made his breakthrough. In one sentence, he wrote he did not invent lithography by "lucky accident" but by "industrious thought," and in another sentence, he said his experiments with mechanical methods on limestone "were entirely checked by a new, accidental discovery" that subsequently led to his invention of the "art" of the purely chemical method of printing. This shows Senefelder was perplexed in his attempt to explain the two opposing determinations that both appeared to him as true. Since he could not have expressed the genuine relationship between accident and necessity in the invention of lithography in a clear and scientific manner, Senefelder instead gave two separate and mutually conflicting explanations for how it happened.

It becomes plain from this that Senefelder himself is responsible for two different stories: one stating that he invented lithography by an "accidental discovery" and another that he arrived at stone printing not "through lucky accident" but by deliberately experimental methods. While this explanation appears to confound rather than clarify matters, Senefelder's contradictory elaboration provides an important clue to solving the riddle of why stories of chance discovery have come to predominate.

The invention of offset printing

Turning now to an examination of the invention of the offset printing method by Ira W. Rubel yields additional clues to answering questions about the peculiar but significant phenomenon of invention by accident. It is a fact that very little has been published about Ira Washington Rubel, the man or the inventor. He was born on August 27, 1860, in Chicago to Moses Rubel, an immigrant from Hochspeyer, Rheinland-Pfalz, Germany, and Ellen (May) Rubel, originally from Philadelphia. He was the eldest of six children, with four brothers, Charles, Simon, Nathaniel, and Levi, and one sister, Bess C. (Rubel) Marks.

Ira Rubel began his working life as a litigating attorney after graduating with a degree in Bachelor of Law from Northwestern University in 1883, and in that same year, he founded, along with his brother, Charles, the Rubel Brothers printing establishment in Chicago. Their printing business thrived and, with the support of two more Rubel brothers, expanded into paper manufacturing and opened an office on Broadway in New York City. Sometime around 1901, the Rubel Brothers Paper Manufacturing Company opened a production facility along the Passaic River in Nutley, New Jersey, and it was at this location that Ira developed the first offset printing press.

Ira died suddenly from a stroke at age forty-eight in 1908 while he was exhibiting his offset press design in Manchester, England. He did not, as far as we know, leave behind a written account of his work as an innovator. The lack of records about Rubel's accomplishment—even though he is universally acknowledged as the inventor of the offset method of printing on paper—has been taken note of by others.

The authors and editors of *The Lithographers Manual* took specific interest in the fact that "the origin of the offset press is one of the least discussed subjects in the literature on printing" and that, "The history of lithography and of the offset press is not yet written." It is also true that major works on the development of printing—for example S. H. Steinberg's *Five Hundred Years of Printing* (1955)—barely mention Ira W. Rubel, offset printing, or the circumstances under which his invention was made. Referring to Rubel as "the America printer" who designed the offset press in 1904 in just one sentence, Steinberg does not repeat the story of accidental invention.

One of the challenges in locating documentary records of Rubel's invention is the fact that he was unable to patent his rubber roller-based offset printing method in the US. If a patent application were available, it is possible details of his work would be in front of us in black and white. However, Rubel was blocked from obtaining a US patent because offset printing on paper was considered by lawyers to be a replication of the tinplate printing method invented in 1875 by Robert Barclay. This technique used cardboard as the "blanket" between the printing plate and the tin substrate. The existing records do show, however, Rubel's inability to patent his invention in the US contributed both to the lack of information about him as well as to his death at a relatively young age. In an obituary published after he died, a family member said his stroke was caused by "the worry and work occasioned in seeking to protect his patents and marketing his inventions in Europe and America."

There are several original sources which explain how the "accidental" attribution came to be applied to Ira W. Rubel's groundbreaking innovation. The editors of *The Lithographers Manual* rely upon an account given by Harry A. Porter, Senior Vice President of the Harris-Seybold Company, in a report to the Detroit Litho Club on December 14, 1950. Porter confirms Rubel operated "a small paper mill in Nutley, New Jersey," where he manufactured "sulfite bond and converted this paper lithographically into bank deposit slips." Significantly, Porter says when Rubel developed the offset press, "lithographic stone presses had a rubber blanket on the surface of their impression cylinder." The impression cylinder "pressed" the paper against the stone and thereby performed the transfer of ink.

The Lithographers Manual goes on, "Whenever the feeder, then not a machine but a person, missed feeding a sheet when the press was operating, the inked image was transferred to the rubber blanket from the stone. The following sheet would then be printed on both sides because the rubber blanket transferred the inked image to the back of the sheet. It was generally known that this unintentionally made transfer produced a print superior to that made directly from the stone. Mr. Rubel noticed this fact and decided to utilize it as the basis of a printing press."

The editors of *The Lithographers Manual* also referenced a description by Frank Heywood in the book by F. T. Corkett, *Photo-Litho and Offset Printing* (1923), which makes clear Rubel was a determined inventor. He took an offset press that he designed to England in 1906, and "The manufacture of this machine was undertaken by a firm of Lancashire engineers, and although for various reasons, the principal being Rubel's somewhat untimely death in 1908, it failed to make good, his efforts must be recognized as beneficent and a distinct contribution to lithographic offset print."

An anecdotal description of the events in Rubel's workshop around 1904—and the accidental way he made his discovery—is provided by Carl Richard Greer in his *Advertising and its Mechanical Production* (1931): "The boy who was feeding the press forgot to send a sheet through, with the result that the image on the stone was transferred, or offset, on the rubber blanket. When the next sheet went through, it did not give the effect Rubel desired, and he threw it aside. The sheet turned over, and on its back, but printed in reverse,

Rubel found the design printed exactly as he desired. He asked the boy how this had happened and was told. For the remainder of the afternoon they experimented, and then Rubel went home and set to work on the design of a press to print indirectly by offset from a rubber blanket."

The Smithsonian Institution possesses at its National Museum of American History in Washington, DC, one of the first presses built by Rubel. The Smithsonian brief describes the machine and the business relations Rubel established with others to develop his design and build presses for sales and distribution in the US. Smithsonian does not make reference to Rubel having arrived at his invention by way of an accidental discovery although it does relate that the press was operated at his facility in New York in 1904 and sold one year later to a printing firm in San Francisco.

The Smithsonian published the following description of Rubel's press in 1996:

> This sheet-fed rotary offset press was built in 1903 by Ira Rubel of Nutley, New Jersey. Its cylinder measures 36 inches in diameter.
>
> The Rubel offset press was the earliest of several rotary, offset machines produced in the first decade of the twentieth century. It was invented in 1903 by Ira Washington Rubel, the owner of a small paper mill and lithographic shop in Nutley, New Jersey. No businessman himself, Rubel formed a partnership early in 1906 with a Chicago lithographer, Alex Sherwood, setting up the Sherbel Syndicate as a monopoly to distribute the press. Sherbel presses were built for the syndicate by the Potter Printing Press Company of Plainfield, New Jersey. The syndicate failed later that year, and the press was redesigned and sold as the Potter offset press, becoming the chief rival to the Harris offset press. Eventually, in 1926, the Potter and Harris companies were consolidated.

Rubel himself went to England to promote his machine in 1907 and died there in 1908 at the age of forty-eight.

This model was operated in Rubel's plant in New York in 1904. In 1905, it was purchased by the Union Lithographic Company of San Francisco for $5,500 and shipped to California. It waited out the San Francisco earthquake and fire on a wharf in Oakland and was put to work in 1907. The maximum speed of the press boasted about 2500 sheets per hour; the sheet size was 28 inches by 34 inches.

Additional information about Ira W. Rubel as an innovator—also minus any reference to misfeeds or accidents—was written by his business associate Frederick W. Sears of New Zealand and published in the *Penrose Pictorial Annual: A Review of the Graphic Arts, Vol. XIV 1908-09*. After the construction of twelve machines and the failure of the Sherbel Syndicate—later giving rise to the Harris-Seybold Company as the primary manufacturer of offset presses in the US—Rubel traveled for the first time to England, as mentioned above, in 1906. It was then that he met and established a relationship with Sears, and the two men agreed to build and sell presses based on Rubel's design in cooperation with the group of Lancashire engineers.

Sears wrote the following tribute to Rubel at the time of his death. It establishes that the man from Nutley, New Jersey, persevered through great difficulties in his work as an innovator and that he was also a fine gentleman:

> There is no doubt, however, that Rubel was the man who showed the world what the offset machine could do, and although there are several makers of these machines today, Rubel's stands out in front of them all. I met him the first day he arrived in England, some three years ago. I

was the first to see his machine run in London, and I joined business with him and was with him to the last. He was the kindest and gentlest-natured man I have ever known, and everyone with whom he came in contact liked or loved him. Some twelve months ago, he had a slight stroke of paralysis at the Derby Hotel, Bury, when we were at tea, but with great care, he pulled round and was able to visit his native land, returning to England in February last. He was never the same man, occasionally he appeared to be himself again, and we all tried to believe the worst was over, but the warning had been given, and our hopes were vain. On Wednesday, the second September 1908, while we were sitting at lunch at the Derby Hotel, the hand of death was laid on him. He dreamily dosed and opened his eyes—we carried him to bed—and he never opened them again. He was conscious only occasionally and died at 9:10 p.m. on the fourth September 1908. His body was cremated at Manchester on the following Monday, and the ashes are to go to his native place to be laid at rest in the family vault beside his father, mother, and brother. Nobody who is not related to him will miss Rubel more than I do. I cannot yet realize that he is no more. I seem to look for him and his letters which came every day.

Ira Rubel's remains are interred at Jewish Graceland Cemetery (Hebrew Benevolent Society Cemetery) in Chicago near a marker that includes both his name and that of his wife, Sarah. It is unfortunate that he did not live to experience the recognition he would later receive the world over for his innovation, and, as far as we know, neither he nor anyone in his family ever benefited financially from the invention.

From the above review, the following can be concluded: (1) Rubel spent years working on the perfection of the offset press design; (2) the "unintentional" transfer of ink to "the back of the sheet" was "generally known" as a technical problem by owners of lithographic presses of that era; (3) Rubel's unique contribution was not only that he experienced this misfeed problem on his rotary lithographic press but that he decided to exploit it; (4) Rubel noticed that the accidentally reversed and indirect image from the rubber blanket of the impression cylinder onto the back of the sheet was superior to that of the right-reading image printed directly from the lithographic printing surface onto the same sheet of paper; and (5) he experimented from this point forward and worked on "the design of a press" based upon the indirect offset method of transferring ink from a lithographic surface to paper.

The importance of the four descriptions of Ira W. Rubel and his invention of the offset printing method—by Porter in 1950 and Heywood in 1923 as published in *The Lithographers Manual,* by Greer in his 1931 book *Advertising and its Mechanical Production,* by the Smithsonian National Museum of American History in 1996, and by Sears as published in 1908 in the *Penrose Pictorial Annual*—is that they establish two important facts about his disruptive advancement that parallel the previous autobiographical description from Senefelder of his invention. On the one hand, something accidental occurred, and then on the other hand, the inventor foresaw the potential contained within this chance event and used it to bring about a significant technological transformation.

In the case of Alois Senefelder, a chance writing of a laundry list upon a limestone revealed possibilities to the determined inventor which he had not previously considered. Further experimentation to exploit the accidentally discovered properties of the grease pencil upon the stone led Senefelder to invent an entirely new printing process based—not upon the mechanical transfer of ink from a raised surface to the paper—but upon the chemically separative properties of oil and water—that is, the ink was attracted to the image on the limestone made with an oil-based writing implement and repelled by the surface covered with water. Senefelder's breakthrough was a critical step in the transition of print technology from the era of handicraft that began with Gutenberg and lasted for more than three centuries into the age of manufacturing that began at the end of the eighteenth and beginning of the nineteenth centuries.

One hundred years later—and coming near the end of the complete mass industrialization of printing—in the case of Ira W. Rubel, a commonly known misfeed error of a sheet of paper on rotary lithographic printing presses with a rubber blanket impression cylinder led to a significant discovery. Rubel transformed this "mistake" into the foundation for a new printing press design. Of course, no one could have known in 1796 or in 1904 how completely the combination of these two breakthroughs would go on to displace the previously dominant letterpress method and transform the entire printing industry in the twentieth century in the form of offset lithography.

Serendipity in the history of innovation

The above review of the work of Alois Senefelder and Ira W. Rubel shows that the stories of innovation by way of chance are both true and untrue. There is a paradox in the explanations of the origin of these key advancements that led to the transformation of the technology of printing. In both cases, an accidental event occurred while the inventors were pursuing technical progress through the methods of exploration and experimentation. In both cases, the inventors picked up on the potential contained in the inadvertent occurrence and pursued it further.

To get a clearer picture of why the accidental element of invention is given prominence in encyclopedia entries and popular reviews on the history of printing—and why they tend to dominate over evidence the inventors' advancements were far from pure luck—it is useful to discuss this phenomenon more broadly since it has been experienced in scientific and technological progress for thousands of years.

It is not difficult to find examples of "accidental inventions." Lists have been published on news and science websites, such as "Ten Accidental Discoveries That Changed the World" or "The Best

Accidental Inventions" or "30 Life-Changing Inventions That Were Totally Accidental: Yes, The World as We Know It Is Predicated on Happenstance." These summaries cover inventions—including the discoveries that led to consumer product brands such as SuperGlue, Teflon, and Vaseline—where scientists and inventors were attempting to solve a problem and came upon their ultimate discoveries by accident. For the most part, these publishers treat the accidental breakthroughs as a novelty or science trivia. They never really get to the nub of the matter or ask the question as to why this phenomenon is so common. However, there are others who have taken a more considered approach to the subject.

For example, in his 1989 book, *Serendipity: Accidental Discoveries in Science*, Royston M. Roberts addresses such occurrences at length by examining more than seventy instances of it. Included in his review are important chance findings, such as the discovery of the New World by Christopher Columbus in 1492, the discovery of oxygen by Joseph Priestley in 1778, and the discovery of evidence of the big bang by Arno Penzias and Robert Wilson in 1964.

Of interest in this review of print technology is Roberts's examination of the first successful photographic process invented by L. J. M. Daguerre in 1838. Daguerre had been experimenting for about five years on a method for permanently capturing an image projected by a camera obscura—a device originally diagrammed by Leonardo da Vinci in 1519 which today might be called a pinhole camera—on plates coated with metallic compounds when he made his accidental discovery.

Roberts describes the event as follows:

> Daguerre prepared plates of highly polished silver-plated copper and exposed them to iodine vapor, which produced a thin layer of silver iodine on the surface. Using the camera obscura, he exposed these plates, producing a faint image. He tried many ways to intensify this image but with little success. One day, he placed an exposed plate, which had only a faint image and which he

> intended to clean and use again, in a cupboard
> containing various chemicals. After several days,
> Daguerre removed the plate and found, to his
> amazement, a strong image on its surface!

In discussing the history of accidental discoveries, Roberts makes a distinction between those inventions that were sought but made accidentally (pseudoserendipity) and the discoveries made by chance without being sought (genuine serendipity). He explains that the term serendipity was coined by Horace Walpole in 1754 after he read about the adventures of "The Three Princes of Serendip." Serendip (or Serendib) is an ancient name for Ceylon, known today as Sri Lanka. Walpole wrote the three princes "were always making discoveries by accidents and sagacity of things which they were not in quest of…," and he came up with the term to describe his own accidental discoveries.

Roberts also says most of the individuals who have been "blessed by serendipity" are not reluctant to admit their good luck. He writes, "They realize, I believe, that serendipity does not diminish the credit due them for their discovery." Far from it, the retelling of the stories of accidental invention in some ways ensure their breakthroughs are never forgotten. In fact, it is this need for a memorable and tellable story that provides an important impulse for the inventors themselves and others who knew them to place an emphasis on the accidental aspect of their innovations.

While not every invention bares the accidental imprimatur, the phenomenon has occurred more frequently in the history of science and invention than many readers may be aware. As Roberts points out in his review, there are dozens of examples of accidental events and outcomes—often observed initially as inconsequential or meaningless—that merged with the persistence of the scientist or inventor and were transformed into intended or unintended discoveries.

One of the most well-known of these stories covered by Roberts is that of the Greek mathematician Archimedes who lived in the third century BC. Archimedes discovered how to measure the volume of an irregularly shaped object by submerging it in water and measuring the amount of liquid displaced by it. The story goes that Archimedes—who had been asked by King Hiero to determine if his crown was made of pure gold—came to his discovery when he saw water run over the top of the tub as he stepped into a public bath in Syracuse. He became so overwhelmed by the excitement of his chance discovery that he ran into the streets completely naked and declared, "Eureka! Eureka!" ("I found it! I found it!")

While there is no written record of the legend of Archimedes' principle—other than its first appearance in an introduction by Vitruvius in his ninth book of architecture some two hundred years after the event was to have happened—the story has staying power and is, at the very least, entertaining and memorable. These features have ensured—despite some critics who assert that Archimedes would never have uttered the word "eureka" at the time—the story has been repeated countless times over the past 2,100 years.

While Roberts does an excellent job of explaining the "eureka moments" in so many examples of accidental innovation, he does not examine the prevalence of these experiences as a necessary phenomenon and a natural but disruptive stage in the continuum from old to new ideas and technology. For Roberts, the accidental is purely accidental and nothing more. Meanwhile, he does not attempt to provide a justification for the overemphasis on the accidental element—and, in many cases, the mythologizing of the event—in popular accounts of significant scientific and technical discoveries.

Isaac Newton and the falling-apple story

The tale of Sir Isaac Newton's discovery of the law of gravitation provides additional information about happenstance as a common experience in technical-scientific achievement and the way it has been used by innovators and historians alike to explain them. Much has been written about Newton's falling-apple legend, and according to research by the Royal Society of London, the source of the tale is now well-known to have been none other than Isaac Newton himself. Although he left no written record of it, the English mathematician, physicist, astronomer, theologian, and author was twenty-three years old in 1666 when the famous "moment" of gravity's discovery was to have taken place.

There are at least four known instances in which Newton—who was born on January 4, 1643, in Lincolnshire and died on March 31, 1727, in the London borough of Kensington—told the anecdote toward the end of his life. The most complete account of Newton telling his story is provided by William Stukeley, a younger scientist from Lincolnshire who befriended the old man and published the first biography of the scientist called *Memoirs of Sir Isaac Newton's Life* in 1752.

The two men met often as members of the Royal Society, and a conversation about the apple took place on April 15, 1726, a year before Newton's death, as Stukeley recorded it in his handwritten memoir:

> After dinner, the weather being warm, we went into the garden & drank thea under the shade of some apple tree; only he & myself. Amid other discourse, he told me, he was just in the same situation as when formerly the notion of gravitation came into his mind. Why sh[oul]d that apple always descend perpendicularly to the ground, thought he to himself; occasion'd by the fall of an apple as he sat in contemplative mood.

Although Stukeley describes the fruit descending "perpendicularly to the ground" and not upon the scientist's head, he confirms that the falling-apple story—in which "the notion of gravitation came into his mind"—was told to him directly by Isaac Newton.

Stukeley goes on to describe how Newton explained the fruit of his contemplation:

> Why sh[oul]d it not go sideways or upward? But constantly to the earth's center? Assuredly the reason is that the earth draws it. There must be a drawing power in matter. And the sum of the

drawing power in the matter of the earth must be in the earth's center, not in any side of the earth.

Therefore, does this apple fall perpendicularly or toward the center? If matter thus draws matter, it must be proportion of its quantity. Therefore, the apple draws the earth as well as the earth draws the apple.

From this account, it appears Newton used the falling-apple tale for two reasons. The first was to explain how he came upon his breakthrough scientific theory of the "drawing power in matter" that later become known as the universal law of gravitation. The second reason was so he could present the basic principles of his theory in a manner that could be understood by nearly anyone. Since the notion that "the apple draws the earth as well as the earth draws the apple" is, to say the least, counterintuitive, Newton was giving a practical example of how "matter thus draws matter" and this power in matter is in "proportion of its quantity."

The historical record contains other contemporary accounts from people with whom Newton shared the apple tale in the final year of his life. One further example to be related here comes from the French philosopher Voltaire who said he was told about the falling fruit by the husband of Newton's niece, John Conduitt. According to Voltaire, Conduitt told him the event had taken place in "the country near Cambridge" where the scientist had withdrawn in 1666 because of the plague.

Commenting recently on the falling-apple legend, Keith Moore, the librarian of the Royal Society, said, "The story was certainly true, but let's say it got better with the telling." Newton's story was thus embellished over time, and the apple "falling upon his head" was most likely added over years of the "better" telling and retelling of the legend. There was a need for both Newton to tell—and his audience to hear—about the theory of universal gravitation, and the apple story became the vehicle through which the new discovery was explained, with some degree of scientific detail presented in popular form.

In both the displacement method of volume measurement discovered by the mathematician Archimedes and the law of gravitation discovered by Newton, it is evident the legends of eureka moments—whether they actually occurred or not—are historically important and there has been a need for such tales to be told. They have helped fix these events in public consciousness by presenting a twist of fate or seemingly inexplicable series of events as a stroke of luck. Through the "magic" of happenstance, people from the era of Newton—who otherwise had no scientific explanation for the source of ideas—could understand complicated concepts that appeared to contradict common sense.

In his introduction to *Serendipity*, Roberts quotes an important comment from Nobel laureate Paul Flory upon receiving the Priestley Medal of the American Chemical Society, "Significant inventions are not mere accidents. The erroneous view [that they are] is widely held, and it is one that the scientific and technical community, unfortunately, has done little to dispel. Happenstance usually plays a part, to be sure, but there is much more to invention than the popular notion of a bolt from the blue. Knowledge in depth and in breadth are virtual prerequisites. Unless the mind is thoroughly charged beforehand, the proverbial spark of genius, if it should manifest itself probably will find nothing to ignite."

While it is, of course, decisive to recognize the role of the individual scientist and his or her desires and intentions—the content of Flory's concept of a mind that is "thoroughly charged" in advance of the accidental moment—there is also the broader technical, social, and even political environment that must also be recognized as foundational to the discovery, without which the individual scientist or inventor could not have been prepared and the serendipitous moment could not have been recognized. We see here—with Flory's reference to "mere accidents" as though they somehow are of entirely secondary importance—an overemphasis on the knowledge, mind, and motivation of the individual innovator as the singular significant factor.

In the case of Newton, did the law of gravitation come to him both "suddenly" and "all at once" with the fall of an apple in a garden in 1666? As explained by F. E. L. Priestley of the University of Toronto English Department in his 1987 essay "Newton and the Apple," "The more vulgar popular versions of the legend feel no qualms about suggesting that the sight of the falling apple suddenly equipped the genius Newton with all he needed to write the *Principia*, apart from minor details." Clearly, aside from the broader societal context within which Newton lived, worked, and thought—he was part of a generation of what were known at the time as philosophers who were preoccupied with the subject of the paths of falling bodies—he could not have had his falling-apple moment or written his groundbreaking 1687 work the *Principia (Mathematical Principles of Natural Philosophy)* which expounded the laws of motion and of universal gravitation.

Priestley goes on to explain that, in the 1660s, there were various theories being discussed regarding "the paths followed by bodies falling toward the center of the earth" along with experiments that are "fully illustrated in detailed accounts of the controversies." The works of Galileo Galilei, Johan Georg Locher, and Johannes Kepler were being studied as the topic was being debated and ideas exchanged. Priestley writes, "Newton himself took part in the discusssions, so he hardly needed to watch an apple to become aware of the problem of the fall of theories of attraction or the law of inverse squares." It would take Newton another twenty years of experimentation and mathematical calculations to complete his work on the law of gravitation.

Furthermore, Priestley explains a critical and decisive element in the historical circumstances surrounding the genesis of Newton's apple tale, that is, the predominating outlook among the scientific and academic community in the early seventeenth century.

Priestley writes:

> It is very difficult, in our own secular and materialist age, to recapture the whole intellectual atmosphere of Newton's world, and it is almost as difficult, in this age of scientific specialization and professionalism, to grasp Newton's attitude toward science—which he called "philosophy."…

> One gets a truer picture of Newton in his his-
> torical context by thinking of him as primarily
> religious and philosophical than by seeing him
> in terms of a modern professional scientist....
> For Newton, spirit is the active substance, mat-
> ter the inactive, inert substance.... The source of
> all motion, of all activity, is the active substance,
> spirit. Both substances are extended in infinite
> space; matter is finitely extended: spirit, at least
> the divine spirit, is infinitely extended.... The
> mathematical, hence rational order, Newton sees
> as evidence of the divine creation of the world
> system, of the divine ordering and dominion
> constantly active in it.

So while Newton and his contemporaries were studying the science of the law-governed movement of bodies, they had a very limited and unscientific understanding of where the impulse for matter
in motion came from. Given that it would be another three hundred
years after Newton sat in his mother's garden to establish by means of
scientific observation that the source of all motion in the universe is
the "big bang," it goes without saying that even less was understood
in the late 1600s about serendipity and the source of revolutionary
scientific and technical ideas. Some have speculated that Newton's
legendary apple story deliberately employs Judeo-Christian symbol-
ism of the biblical forbidden fruit from the garden of Eden. It was
far more likely—whether the apple did fall on his head or not—that
Newton saw his own discovery as a moment of divine inspiration
rather than merely a stroke of good luck.

Accident, chance, and necessity

From the above review of the prevalence of accidental discovery in technical and scientific progress, it is evident that there are
two sides to this phenomenon. An unexpected or inadvertent mistake or result occurs that becomes, through the engagement of an

inventor, the basis for an important breakthrough. Awareness of this contradictory reality—the ability to understand that all evolutionary processes move simultaneously through deterministic and random elements and, likewise, the steps in technological advancement proceed along a law-governed path which manifest themselves by way of unanticipated particulars—is the touchstone for grasping one of the primary and essential contradictions embedded in the disruptive continuity of human technical progress.

Dictionary definitions of the term "accident" present the concept as an event that is unforeseen or unplanned. While an accident is often associated with negative outcomes (*I was in an auto accident*), it is also frequently used to connote good fortune (*We met by accident*). Although there is a tendency to see these determinations as mutually exclusive—as either purely desirable or undesirable—these opposite meanings of the word express one side of the objective duality of accidents.

Whether accidents are associated with negative or positive results, they always appear to be purely unintended. However, it becomes clear upon examination that an accident was preceded by circumstances causing it to happen, and the accident itself is the cause of still further apparent unanticipated results. Thus, every accident is both cause and effect and expresses both something entirely expected and unexpected at the same time. This is the real and true dialectical nature of the phenomenon.

The term "chance" is defined in a similar manner as an unpredictable event without discernible human intention or observable cause. The existence of a chance happening is conventionally understood to arise at random and without a purpose. However, while the outcome of a chance roll of dice may be thought of by the roller as either bringing "good luck" or "bad luck," a combination of scientifically observable and measurable physical factors—gravity, air resistance, friction, the force of the throw, and the initial position of the dice—interact to produce one of thirty-six possible landing combinations of the two cubes.

Definitions of the term "necessity" emphasize the concept as a quality or state of being that is required or a consequence that could not have turned out any other way. From this understanding, it appears that necessity is predetermined with no influence from accidental elements. In this formal way of thinking, accidents or chance events and necessary events are conceived of as fixed opposites which never touch or interact with each other.

Accounts of accidental invention most often present the moment of discovery by way of happenstance in a similarly one-sided manner. The unscientific perspective erases the contradictory content of an event that is both accidental and necessary and adopts one of its two forms exclusively: either presenting the accidental occurrence as completely disconnected from necessity or by dismissing the accident and overemphasizing the trajectory of an inevitable outcome as the only possible path of development.

In the first instance, the event is seen as itself being purely accidental. This means the accident is thought to express only inconsequential or irrelevant trends. From this point of view, the discovery or invention at a specific moment in time did not come about because it was necessary at all but instead took place purely at random.

In the second instance, focusing on the importance of the motivations of the inventor or attributing the event to "good luck," for example, this standpoint dismisses the accidental element entirely and sees only the necessary or predetermined eventuality.

However, there is a more precise and objectively accurate way of understanding the occurrence of accidents in connection with technical progress that embraces the contradictory relationship between chance and necessity. Instead of viewing an accident like a lifeless cartoon, it can be understood scientifically as the real means through which advancement takes place. Like a 3D movie, there are concatenations of circumstances surrounding and forming the lead up to the moment of a chance discovery. These conditions are the foundation upon which it is possible for the innovator to see the potential contained within the unanticipated event. It is precisely here that the accidental and the necessary merge together and become identical; the accidental is necessary and the necessary is accidental.

This manner of understanding rejects the notion that an invention is either purely a chance event or purely a predestined outcome and instead embraces the view that these two states exist as one and this is the way innovation actually happens. A necessary event has a cause because it is accidental and has no cause because it is accidental; the necessary manifests itself as chance, and this chance also is the expression of necessity.

Among the first thinkers to scientifically define the relationship between chance and necessity in both nature and human thought was the German philosopher Georg Wilhelm Friedrich Hegel (1770–1831). In his *Science of Logic*, Hegel discusses at length the process by which necessity is manifested in phases through the possible, the accidental, and the actual:

> Frequently, nature, to take it first, has been chiefly admired for the richness and variety of its structures…the chequered scene presented by the several varieties of animals and plants, conditioned as it is by outward circumstances—the complex changes in configuration and grouping of clouds and the like—ought not to be ranked higher than the equally casual fancies of the mind which surrenders itself to its own caprices. The wonderment with which such phenomena are welcomed is a most abstract frame of mind from which one should advance to a closer insight into the inner harmony and uniformity of nature….
>
> In respect of mind and its works, just as in the case of nature, we must guard against being so far misled by a well-meant endeavor after rational knowledge as to try to exhibit the necessity of phenomena which are marked by a decided contingency, or, as the phrase is, to construe them *a priori*. Thus, in language (although it be, as it were, the body of thought), chance still unquestionably plays a decided part, and the same is true of the creations of law, of art, etc.…

> Necessity is often said to be blind. If that means that in the process of necessity the end or final cause is not explicitly and overtly present, the statement is correct. The process of necessity begins with the existence of scattered circumstances which appear to have no interconnection and no concern one with another. These circumstances are an immediate actuality which collapses, and out of this negation, a new actuality proceeds....
>
> Necessity is blind only so long as it is not understood.

In a manner similar to that of Newton, Hegel saw all motion and development in nature as being driven by an almighty spirit he called the Absolute Idea. He saw the interaction of chance and necessity in the universe as part of the process through which the Absolute Idea manifested itself in nature. Although Hegel inverted the genuine relationship between thinking and being, he did show how opposite determinations, such as chance and necessity, are united and dependent upon each other in the unfolding process of all matter in motion including in the thought reflections of human consciousness.

The biological evolutionist Charles Darwin (1809–1882) was among the first scientists to work with similar concepts in a specific study of nature. Darwin showed that the process of natural selection—which governs the evolution of biological life and is the continuum of descendance from one creature to the next over the course of 3.7 billion years—takes place through the complex interaction of chance and necessity. Francisco J. Ayala, the noted evolutionary biologist and professor at the University of California, Irvine, wrote that Darwin was the first to uncover and explain the dialectical interdependence of random and deterministic forces at work in the evolutionary progression of life-forms in the natural world. Ayala wrote:

> Natural selection accounts for the "design" of organisms because adaptive variations tend to

increase the probability of survival and reproduction of their carriers at the expense of mal-
adaptive, or less adaptive, variations. The arguments of Aquinas or Paley against the incredible
improbability of chance accounts of the origin of
organisms are well taken as far as they go. But
neither these scholars, nor any other authors
before Darwin, were able to discern that there is
a natural process (namely, natural selection) that
is not random but rather is oriented and able to
generate order or *create*. The traits that organisms
acquire in their evolutionary histories are not fortuitous but determined by their functional utility
to the organisms.

Chance is, nevertheless, an integral part of
the evolutionary process. The mutations that
yield the hereditary variations available to natural selection arise at random, independently of
whether they are beneficial or harmful to their
carriers. But this random process (as well as others that come to play in the great theater of life) is
counteracted by natural selection, which preserves
what is useful and eliminates the harmful....

The theory of evolution manifests chance
and necessity jointly intricated in the stuff of
life; randomness and determinism interlocked
in a natural process that has spurted the most
complex, diverse, and beautiful entities in the
universe: the organisms that populate the earth,
including humans who think and love, endowed
with free will and creative powers, and able
to analyze the process of evolution itself that
brought them into existence.

As Ayala explains, not every random mutation or evolutionary trait contributes to the viability of an organism. Some prove to

be harmful. Nonetheless, these seemingly negative random events are themselves also driven by the variety, complexity, and interconnectedness of the material circumstances contributing to biological evolution. Here, it is evident that what proves to be unnecessary in nature is also at the same time the product of necessity.

There were still other great thinkers and men of science in the nineteenth and twentieth centuries who, basing themselves upon the work of Hegel and Darwin, arrived at a more complete and scientifically accurate understanding of the relationship between chance and necessity as it applies to all of nature, including the active side of thought in the form of historically determined social practice.

For example, the philosopher and economist Karl Marx endorsed the Hegelian dialectic of evolutionary development through contradiction and the unity of opposites such as chance and necessity. At the same time, Marx corrected Hegel's inverted superior relationship of consciousness to nature and restored the primacy of material reality as the source of the thought reflections in the human mind, while also accepting the fundamentally creative and practical side of thinking as a social and historical process. On the specific question of the evolution of technology, Marx discussed the development of what he referred to as the material forces of production extensively in his writings and used the term *technology* in connection with his analysis of the transition from the era of handicraft tools to that of powered machines within the capitalist system. In identifying the beginning of the industrial revolution with John Wyatt's spinning machine of 1735, Marx wrote an important footnote in his well-known volume, *Capital*:

> A critical history of technology would show how little any of the inventions of the 18th century are the work of a single individual. Hitherto there is no such book. Darwin has interested us in the history of Nature's Technology, i.e., in the formation of the organs of plants and animals, which organs serve as instruments of production for sustaining life. Does not the history of the

productive organs of man, of organs that are the
material basis of all social organization, deserve
equal attention? And would not such a history
be easier to compile, since, as Vico says, human
history differs from natural history in this, that
we have made the former, but not the latter?
Technology discloses man's mode of dealing with
Nature, the process of production by which he
sustains his life, and thereby also lays bare the
mode of formation of his social relations, and of
the mental conceptions that flow from them.

From the viewpoint of Marx and other adherents to historical materialism, a particular breakthrough event becomes a necessity
by way of socioeconomic development, and the specific individual
who emerges during the course of the achievement is the result of a
complex set of objective, subjective, deterministic and accidental circumstances. As is clear from this study of the history printing—and
presumably any study of technology development in the era of industrial machines—the work of innovation has become less the result
of individual achievement and more a social process involving many
individuals or groups of people connected to one another through
the structures of economic activity and development. With this key
aspect of technological disruptive continuity in mind, one can see
that the engine of progress drives the whole of society forward, and if
one individual had not stepped forward, the necessary advancement
would, in the long run, have been brought about by someone else.

These laws of development apply to scientific and technical
advancement not as predestined outcomes, but as accomplishments
along a continuum that is full of random events that are punctuated
by the additional decisive variable of the consciousness of human par-
ticipants. This means that, regardless of what individual innovators
may think about their activity or understand about the significance
of the discoveries they are making, there is a logic through which
their advancements proceed. As with natural evolutionary processes,
the movement of technical innovation—including the ideas devel-

oped by those who contribute to the progress attained—follows a discernible path that can be examined, traced, and understood.

Like the biological evolutionary advancement from simple to complex life-forms and from unconscious to conscious beings—after passing through a vast number of seemingly accidental physical traits and characteristics—socioeconomic technical progress has gone from primitive stone tools over generations to the most sophisticated and complex systems that enable the exploration of the surface of Mars and the examination of the outer reaches of the solar system and beyond. While each of the breakthroughs along this path have come about in apparent randomness—including both the individual person or people responsible for them as well as the way the innovations were secured—the advancements have proceeded along a logical series of steps that track with the stages of socioeconomic history and the philosophical and scientific ideas that emerged as the crowning achievements of those stages.

Disruptive continuity and the history of printing

To return to Alois Senefelder and Ira W. Rubel, it can be stated that these inventors were responsible for two critical and necessary stages in the development of print technology. In the first case, Senefelder was responsible for the transition from the relief-mechanical to the planographic-chemical era of printing. This means the process developed by Senefelder no longer relied upon the mechanical impression of ink from raised letterforms into the paper but upon the chemically separative properties of oil and water on an otherwise flat surface of lithographic stone for transferring a printing image to paper.

In the second case, Rubel was responsible for the transition from the direct image transfer to the indirect image transfer stage of printing-press technology. This means Rubel redesigned the printing press to include an additional rubber cylinder mechanism that stood between the inked planographic surface from which it accepted the printing image and transferred it to the paper with the assistance of an impression cylinder. Together, these two advancements established offset lithography, the high-speed, mass manufacturing method that overtook letterpress printing and became dominant internationally, especially in the second half of the twentieth century.

These two transformations first manifested themselves in the form of accidental events: Senefelder's chance writing of a laundry list upon a limestone with a crayon and Rubel's paper misfeed on a lithographic rotary printing press with a rubber blanket on its impression cylinder. As discussed above, that the breakthrough to offset lithography is associated with accidents is entirely consistent with the manner which human technical progress takes place. The combination of scientific and socioeconomic changes that occurred in the eighteenth and nineteenth centuries made necessary a departure of print technology away from the letterpress printing processes and into entirely new planographic, chemical, and indirect printing processes of the modern industrial era. While the two innovators were working to develop printing methods in their respective lifetimes, the necessary departure from the system associated with Johannes Gutenberg—preconditioned by more than three centuries of societal change—appeared initially as chance events to both Senefelder and Rubel.

If these two individual inventors had not made these breakthroughs, would offset lithography have been invented by others, would this transformation still have happened sooner or later? Had Senefelder been content to become a playwright and Rubel stayed in the papermaking business, would the logic of industrial society and the need for a high-speed and mass manufacturing method of print communications in the twentieth century still have resulted in offset lithography? When looking at Senefelder and Rubel within a broader historical context, an affirmative answer must be given to these counterfactual questions.

The role and interaction of the macro socioeconomic (objective) and the micro individual inventor (subjective) driving forces of advancement are accounted for within the theoretical framework of disruptive continuity. Beginning with the societal context, the theory first posits that powerful evolutionary trends in human culture bind events together in a historical progression, from lower to higher and from simple to more complex forms. This progression of technology is intimately bound up with the stage of social organization and intellectual development of mankind. This is not to say that there have never been reversals of progress or that every technological innovation is a guaranteed contribution to further advancement. As is the case with biological evolution, where some mutations produce traits that are harmful to the survival of a species, some technical innovations have, under certain conditions, represented a step backward or a diversion along a dead-end path.

There is also the matter of technical advancements becoming the means through which society can be destroyed as, for example, would happen in the event of a nuclear world war or in earlier societies that collapsed from the loss of soil fertility due to crop growing and farming practices. This danger arises from a conflict between the increasing sophistication and power of human innovation with the inability of the cultural level and social and political organization of civilization to adequately accommodate the technological advancements. However, despite these periodic setbacks and existential threats, the general dynamic of development is one of an ever more complex and greater technical subordination of the properties and laws of nature to the needs of man and an expanding separation of humanity from the blind operation of forces both within the natural environment and society itself, which is also in the end a product of nature.

The topic of the continuous versus the discontinuous view and the role of the individual inventor in the development of man-made implements and artifacts has been the subject of study by historians. George Basalla, a historian of science at the University of Delaware, deals with this question in his book *The Evolution of Technology*. Basalla likens the history of technology to biological evolution and correctly objects to those who advocate the view that "inventions are the products of superior persons who owe little or nothing to the past." He also separates his own position from those who support a theory of a purely discontinuous series of "scientific revolutions" with "more sophisticated formulations" rendering all technical innovation into discrete phases, each with no relationship to prior accomplishments whatsoever.

Basalla says that technology should not be seen in a linear relationship with science and placed in a "subordinate position" to the latter and with the former "erroneously defined as the application of scientific theory to the solution of practical problems." He writes, "Technology is not the servant of science" because it existed long before science, and "people continued to produce technical triumphs that did not draw upon theoretical knowledge."

However, Basalla goes on to bend the stick back too far in the opposite direction, writing:

> The artifact—not scientific knowledge, not the technical community, not social and economic factors—is central to technology and technological change. Although science and technology both involve cognitive processes, their end results are not the same. The final product of innovative scientific activity is most likely a written statement, the scientific paper, announcing an experimental finding or a new theoretical position. By contrast, the final product of innovative technological activity is typically an addition to the made world: a stone hammer, a clock, an electric motor.

Instead of recognizing that the growth of knowledge and science developed, at first, in tandem with the early technological development of society's productive forces and then rose above them in the form of theoretical science and applied research, Basalla inverts the ahistorical demotion of the product of innovation and says that the artifact stands head and shoulders above science. In examining the impact of solid-state physics on the development of semiconductors, for example, the essential role of basic science on the further advancement of a core technology of the electronic age becomes clear. Without ongoing fundamental research into materials science, the accelerating miniaturization, reliability, speed and energy efficiency behind Moore's law—the theorem of progress that says the number of transistors on a microchip doubles every two years—would be impossible.

Basalla sums up his theory of technology evolution by referring to the historian Brooke Hindle who he says argued that "the artifact in technology" is superior to any "intellectual or social pursuits" because it is a "product of the human intellect and imagination and, as with any work of art, can never be adequately replaced by a verbal description." These are peculiar ideas being advanced by someone who is opposing the standpoint of the theory of technology evolution by exclusive successive stages of discontinuity. While Basalla insists upon the existence of an uninterrupted and perpetual continuity in the products of human technical innovation, his presentation is the polar opposite of that which he criticizes and is, therefore, also marred by a similar if not identical one-sidedness. The dichotomy between science and technology, between the individual artifact and social context, between discontinuity and continuity in human innovation, and between the practical and theoretical sides of invention are all conceived of as fixed categories that never touch each other, are incapable of interpenetrating and cannot be understood in thought as existing simultaneously as part of a historically developing process. In Basalla's concept of technical innovation, there are no breaks or disjunctions in human progress and, instead, technical history moves along ever so smoothly, morphing from one object to the next without any abrupt dislocations or transformations.

The preoccupation with continuity to the absolute exclusion of discontinuity and singular focus on the primacy of the individual artifact above scientific theory means that one cannot account for the living relationship between disruptive advancements in society and how they drive revolutionary leaps in technology. In the search to find evidence of antecedent developments for every innovation in human history—such as stone tools, the cotton gin, steam power, the electric motor, the light bulb, etc.—Basalla is forced to suppress the actual discontinuity that occurs in the progression from one innovation to the next. In the end, however, Basalla betrays his initial statement against the role of the "superior persons who owe little or nothing to the past" by embracing the idea that the "human intellect and imagination" and "work of art" are the penultimate expression of technical progress.

Even if it were possible to trace every object of human ingenuity backward in time, there would be a point at which one would arrive at the first man-made thing ever created. Then like the chicken-or-egg dilemma, how does the idea that every artifact ever created by man has a precursor stand up? Outside of a recognition that there was a simultaneously continuous and discontinuous vault in man's earliest development—a phenomenon repeated throughout history—the first truly human tool would never have been created and technical progress would never have gotten started. Thus, the theory of a purely continuous evolutionary development of technology does not withstand scrutiny.

Disruptive continuity, on the other hand, is a theory within which it is possible to understand both the interconnection of every invention with the past as well as its separation and elevation above the limits of prior milestones, driven by the increasingly complex accumulation of man's productive capacities, social organization, and intellectual accomplishments. Given the interrelationship of human communications—and especially print communications—with the practical, technical, scientific and philosophical development of society, there is perhaps no better sphere of historical and technical analysis that substantiates the validity of the theory of disruptive continuity.

The socio-economics of human communications

All forms of human communication have existed as both by-products of and engines for the advancement of social existence from the earliest humans (homo sapiens) to the first forms of civilization to the modern information age. During the early Stone Age, when speech and language—the first truly human communication—began somewhere between 1.75 and 3 million years ago, prehistoric man in the Lower Paleolithic was crafting Acheulean stone axes, coordinating hunting and gathering and engaging in other complex social practices that depended upon verbal communications. While these primitive peoples had speech, they had no repository for recording their thoughts or words.

As Harvey Levenson of California Polytechnic State University explains in *Understanding Graphic Communications*, "Early peoples left little information about themselves, as they had no way to transform spoken words into written languages, to pass on their heritage to future generations. Their history and their wisdom died with them. It was not until fairly recently in human history that people figured out ways to record simple events and ideas through pictures and symbols."

The earliest visual communication, such as the cave paintings at Lascaux in southern France, were created by anatomically modern humans somewhere between seventeen thousand and thirty-five thousand years ago during the Upper Paleolithic (late Stone Age). By this point, humans had migrated to all inhabitable areas of the earth, and these tribal peoples were engaged in what is known as "behavioral modernity." The use of pictorial communication proceeded along with abstract thinking, planning depth, fishing, burial practices, ornamentation, music, dance, and other cultural activities connected with more advanced blade technology and toolmaking practices.

Although we do not precisely know the meaning of the pictures at Lascaux Cave, the nearly six thousand painted figures are representations of the animals—including bulls, equines, stags, and a bear—that existed in Europe during that era. The paintings also include representations of humans and other abstract symbols indicating a high degree of self-awareness and ingenuity. Significantly, the people who made the cave paintings used scaffolding to reach the ceilings and created the colors of red, yellow, and black from a wide range of mineral pigments including compounds such as iron oxide and substances containing manganese. In some instances, the color is thought to have been applied by suspending the pigment in either animal fat or clay as a primitive paint. The colors were swabbed or blotted on, and it appears that the pigment was applied by blowing a mixture through a tube. All these facts help to build an understanding of the way the technical level of society corresponds with the development of increasingly complex human communications from strictly verbal to varieties of visual and pictorial forms.

There are no records of who the individuals were that uttered the very first words of speech or the first sentence containing a noun and a verb or who painted the first picture on a cave wall or who developed the first paint used in cave art. While the location of these occurrences and the anonymous people involved came about in a seemingly random fashion, there is no doubt these advancements were conditioned by objective material circumstances and became necessary steps in the development of human culture. They were both a contribution to and the consequence of the social and technical environment of the early and late Stone Ages.

Similar connections can be deduced from the stamp seals used by Sumerian scribes to identify ownership in Mesopotamia in 4500 BC—the earliest forms of writing called cuneiform—who employed pictographs to communicate syllables and sounds that were later pressed into clay with a wedge-tipped stylus in 3300 BC. These relationships can also be seen in the hieroglyphics written on papyrus, a precursor to paper, developed in Egypt in 3200 BC. All these advancements in graphic communications—developed in an apparent unplanned and haphazard manner—are rooted in the socioeconomics of ancient civilization, especially the expansion of commerce and the need for record keeping. These events represent critical chapters in the prehistory of print communications.

In examining the history of printing, it is rare to find surveys that trace the interactions between the evolution of the broader technical foundations of society—punctuated by the unpredictable randomness of inventors and the means through which they arrived at their inventions—with the transitions in the means and methods of print communications. This problem was highlighted by Elizabeth L. Eisenstein in her 1979 volume *The Printing Press as the Agent of Change*. Eisenstein explains that the "far-reaching effects" of Gutenberg's invention—which "left no field of human enterprise untouched" and whose consequences are "of major historical significance"—have so far seen very little elaboration in the major texts on the history of print technology. Eisenstein points out that historians may refer to printing as one of the most important inventions, if not the most important invention, of the second millennium, but the dynamic interrelationship of Gutenberg's breakthrough—as well as subsequent advancements—with the broader processes of human progress are rarely explored. As Eisenstein wrote in her essential work on the societal causes and impacts of printing:

Insofar as flesh-and-blood historians who turn out articles and books actually bear witness to what happened in the past, the effect on society of the development of printing, far from appearing cataclysmic, is remarkably inconspicuous. Many studies of developments during the last five centuries say nothing about it at all.

Those who do touch on the topic usually agree that the use of the invention had far-reaching effects. Francis Bacon's aphorism suggesting that it changed "the appearance of and state of the whole world" is cited repeatedly and with approbation. But although many scholars concur with Bacon's opinion, very few have tried to follow his advice and "take note of the force, effect, and consequences" of Gutenberg's invention. Many efforts have been made to define just what Gutenberg did "invent," to describe how movable type was first utilized and how the use of the new presses spread. But almost no studies devoted to the consequences that ensued once printers had begun to ply their new trades throughout Europe. Explicit theories as to what these consequences were have not yet been proposed, let alone tested or contested.

The lack of research that Eisenstein characterizes in her book as "The Unacknowledged Revolution" is due in part to a frequent presentation of print-technology history in a linear manner, going from one process to another—or from one innovator to another—in isolation from an analysis or understanding of the broader socioeconomic context and the real driving forces of change before and since 1440.

It is now well-established that printing was first developed in China, where paper was invented under the Han Dynasty by Ts'ai Lun (Cai Lun) at the beginning of the second century. The first woodblock printing known as xylography—where a complete page of relief characters was carved in a piece of wood—was produced in China in the sixth century. It is also known that movable typeforms made of clay, ceramic, wood, and metal were also developed first in China between the eleventh and thirteenth centuries. Some of these methods spread across East Asia, including Korea and Japan, and into the western regions of Central Asia. It is speculated but not confirmed that the Asian relief printing techniques made their way into Europe where they were developed further in the form of the handheld metal-casting mold and mechanical printing press invented by Johannes Gutenberg in the fifteenth century. It is known that Gutenberg's typography methods were reintroduced to China in the nineteenth century.

The reason the birth of printing has become associated with Gutenberg and his invention in Mainz, Germany, in 1440—and not in China at least three hundred years earlier—is that the entire mechanized production system created by the fifteenth-century inventor was vastly more productive than the hand-carving xylographic techniques from the East. Due to a series of other foundational events taking place at the time, Gutenberg's methods were picked up by a cultural transformation underway in Europe and swept rapidly from one country to another. Meanwhile, the movable-type techniques first pioneered in Asia did not overtake woodblock printing the way it did in the West. This was at least partially because Chinese interchangeable-type production required the management of as many as forty-thousand different characters. While xylography existed in Europe and was carried over into Gutenberg's time for illumination of printed texts from the earlier form of handwritten manuscripts, the methods he developed represented a paradigm shift away from all previous techniques up to that point in world history. Eventually, woodblock illustrations were replaced by other methods of pictorial reproduction that were derivative and complementary to the relief process and printing-press mechanism of the letterpress era.

Perhaps the best description of the reciprocal impact of Gutenberg's invention is provided by Will Durant in the sixth volume of *The Story of Civilization—The Reformation*:

Soon half the European population was reading as never before, and a passion for books became one of the effervescent ingredients of the Reformation age.... The typographical revolution was on.

To describe all its effects would be to chronicle half the history of the modern mind.... Printing replaced esoteric manuscripts with inexpensive texts rapidly multiplied, in copies more exact and legible than before, and so uniform that scholars in diverse countries could work with one another by references to specific pages of specific editions.... Printing published—i.e., made available to the public—cheap manuals of instruction in religion, literature, history, and science; it became the greatest and cheapest of all universities, open to all. It did not produce the Renaissance, but it paved the way for the Enlightenment, for the American and French Revolutions, for democracy. It made the Bible a common possession and prepared the people for Luther's appeal from the popes to the gospels; later it would permit the rationalist's appeal from the gospels to reason. It ended the clerical monopoly of learning, the priestly control of education. It encouraged vernacular literatures for the large audience it required could not be reached through Latin. It facilitated the international communication and cooperation of scientists. It affected the quality and character of literature by subjecting authors to the purse and taste of the middle classes rather than to aristocratic or ecclesiastical patrons. And, after speech, it provided a readier instrument for the dissemination of nonsense than the world has ever known until our time.

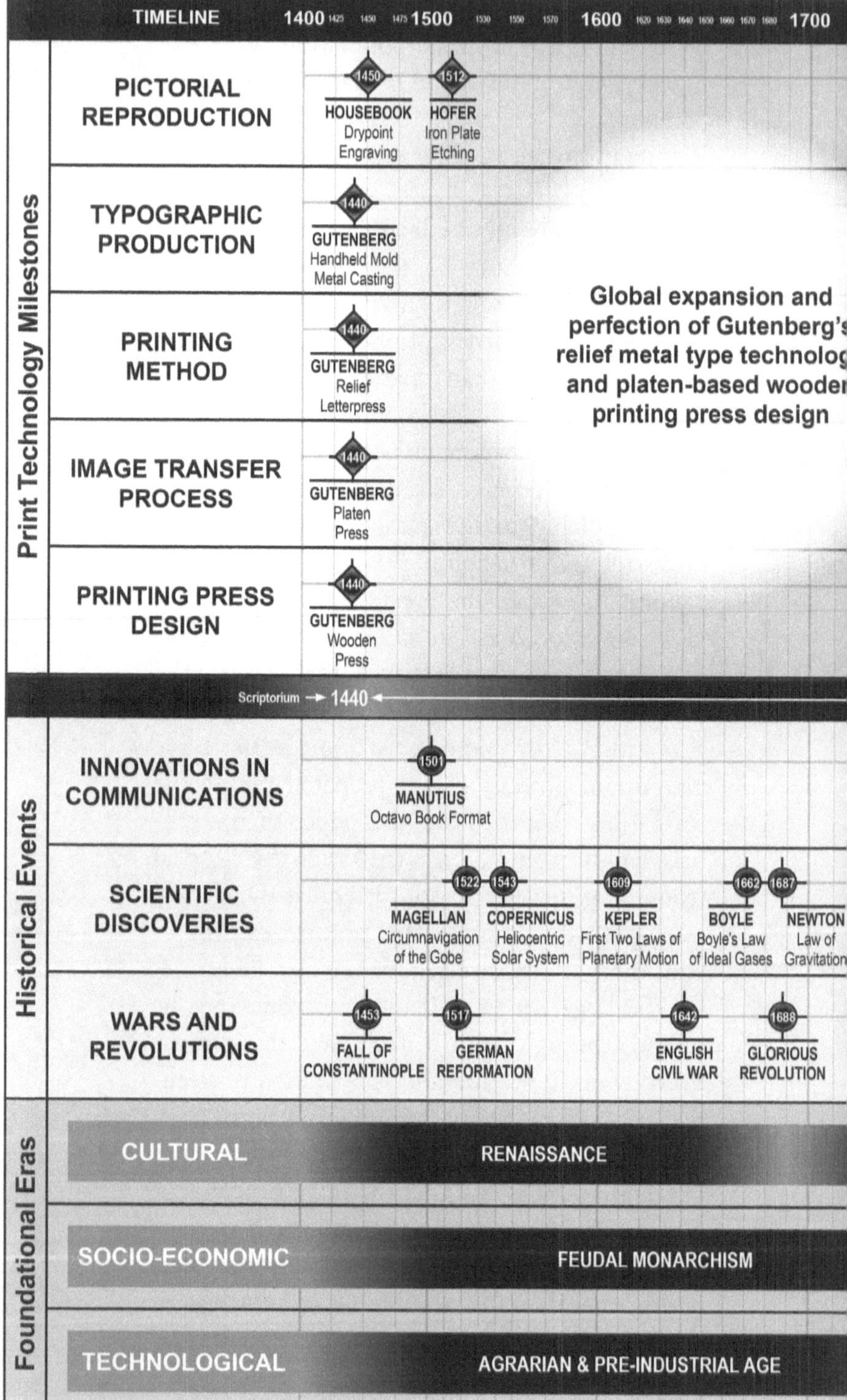
TIMELINE
1400 1425 1450 1475 1500 1530 1550 1570 1600 1620 1630 1640 1650 1660 1670 1680 1700
Print Technology Milestones
PICTORIAL REPRODUCTION
1450
HOUSEBOOK
Drypoint Engraving
1512
HOFER
Iron Plate Etching
TYPOGRAPHIC PRODUCTION
1440
GUTENBERG
Handheld Mold Metal Casting
PRINTING METHOD
1440
GUTENBERG
Relief Letterpress
IMAGE TRANSFER PROCESS
1440
GUTENBERG
Platen Press
PRINTING PRESS DESIGN
1440
GUTENBERG
Wooden Press
Global expansion and perfection of Gutenberg's relief metal type technology and platen-based wooden printing press design
Scriptorium 1440
Historical Events
INNOVATIONS IN COMMUNICATIONS
1501
MANUTIUS
Octavo Book Format
SCIENTIFIC DISCOVERIES
1522
MAGELLAN
Circumnavigation of the Gobe
1543
COPERNICUS
Heliocentric Solar System
1609
KEPLER
First Two Laws of Planetary Motion
1662
BOYLE
Boyle's Law of Ideal Gases
1687
NEWTON
Law of Gravitation
WARS AND REVOLUTIONS
1453
FALL OF CONSTANTINOPLE
1517
GERMAN REFORMATION
1642
ENGLISH CIVIL WAR
1688
GLORIOUS REVOLUTION
Foundational Eras
CULTURAL
RENAISSANCE
SOCIO-ECONOMIC
FEUDAL MONARCHISM
TECHNOLOGICAL
AGRARIAN & PRE-INDUSTRIAL AGE

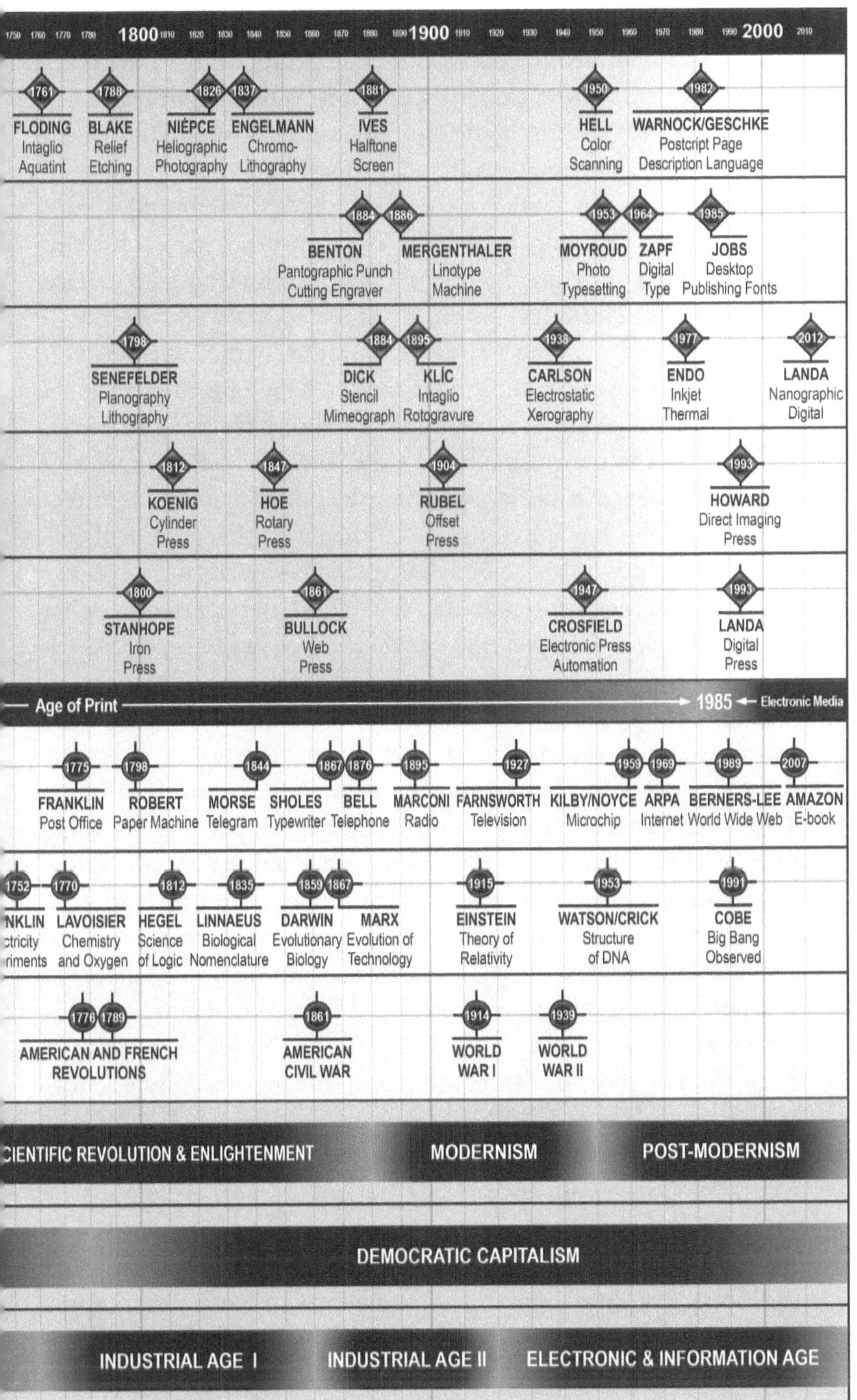
1750 1760 1770 1780 1800 1810 1820 1830 1840 1850 1860 1870 1880 1890 1900 1910 1920 1930 1940 1950 1960 1970 1980 1990 2000 2010
1761 FLODING Intaglio Aquatint
1788 BLAKE Relief Etching
1826 NIÉPCE Heliographic Photography
1837 ENGELMANN Chromo-Lithography
1881 IVES Halftone Screen
1950 HELL Color Scanning
1982 WARNOCK/GESCHKE Postcript Page Description Language
1884 BENTON Pantographic Punch Cutting Engraver
1886 MERGENTHALER Linotype Machine
1953 MOYROUD Photo Typesetting
1964 ZAPF Digital Type
1985 JOBS Desktop Publishing Fonts
1798 SENEFELDER Planography Lithography
1884 DICK Stencil Mimeograph
1895 KLÍC Intaglio Rotogravure
1938 CARLSON Electrostatic Xerography
1977 ENDO Inkjet Thermal
2012 LANDA Nanographic Digital
1812 KOENIG Cylinder Press
1847 HOE Rotary Press
1904 RUBEL Offset Press
1993 HOWARD Direct Imaging Press
1800 STANHOPE Iron Press
1861 BULLOCK Web Press
1947 CROSFIELD Electronic Press Automation
1993 LANDA Digital Press
Age of Print
1985 Electronic Media
1775 FRANKLIN Post Office
1798 ROBERT Paper Machine
1844 MORSE Telegram
1867 SHOLES Typewriter
1876 BELL Telephone
1895 MARCONI Radio
1927 FARNSWORTH Television
1959 KILBY/NOYCE Microchip
1969 ARPA Internet
1989 BERNERS-LEE World Wide Web
2007 AMAZON E-book
1752 FRANKLIN Electricity Experiments
1770 LAVOISIER Chemistry and Oxygen
1812 HEGEL Science of Logic
1835 LINNAEUS Biological Nomenclature
1859 DARWIN Evolutionary Biology
1867 MARX Evolution of Technology
1915 EINSTEIN Theory of Relativity
1953 WATSON/CRICK Structure of DNA
1991 COBE Big Bang Observed
1776 1789 AMERICAN AND FRENCH REVOLUTIONS
1861 AMERICAN CIVIL WAR
1914 WORLD WAR I
1939 WORLD WAR II
SCIENTIFIC REVOLUTION & ENLIGHTENMENT
MODERNISM
POST-MODERNISM
DEMOCRATIC CAPITALISM
INDUSTRIAL AGE I
INDUSTRIAL AGE II
ELECTRONIC & INFORMATION AGE

The above timeline illustrates the interrelationship of the milestones in printing innovation between the fifteenth and beginning of the twenty-first centuries with the progression of the foundational eras of society—technological, socioeconomic, and cultural—and the major events in world history, science, and communications. This visual presentation establishes a framework within which to understand the unfolding process of disruptive continuity in the age of print communications.

The birth of printing was both the product of the increasing volume of hand copying by the scribes and the development of artisanal methods of handicraft metallurgy associated with the transition from the Dark Ages to the great cultural awakening of the Renaissance. It was a consequence of and a catalyst for the flowering of art, architecture, philosophy, literature, music, politics, and other aspects of culture that were driven by the broad progression of societal change that brought about the discovery of the New World, the circumnavigation of the globe, and the opening of trade routes from Europe through Asia and America. The printing press and the associated circulation of printed books and other materials to ever wider layers of the population were instrumental in great social movements, such as that of the Protestant Reformation sparked by Martin Luther's critique of the indulgences of the Catholic Church. Behind these changes came the Enlightenment of the seventeenth and eighteenth centuries, the development of democratic government, the domination of the commodity economy, and the supercession of the institutions of feudalism during the American and French Revolutions. One need only point to the significance of the publication of Thomas Paine's *The Age of Reason* to illustrate the fundamental role played by print in the intellectual ferment that accompanied the transition from feudal absolutism to democratic capitalism.

The systems of the printing press continued to exist as they had been developed by Gutenberg—based upon fruit presses designed for making cider, wine, and oils—for three and a half centuries before any significant design changes were introduced. The technology developed very slowly like in an incubator with few modifications, surrounded by the steady development of the Renaissance until the rise of the Scientific Revolution and the coming of the first industrial revolution. Then new materials and mechanical techniques exploded the old wooden machine from Gutenberg's era, and by the beginning of the nineteenth century, the printing press evolved rapidly into a completely iron apparatus with levers (Stanhope, 1800), replacing the physical strength of pressmen needed to operate the old screw mechanism to transfer ink onto paper. At the same time, the image transfer technologies evolved first into a metal cylinder-to-platen hybrid (Koenig and Bauer, 1818) and then, by the middle of the 1800s, to a fully cylinder-based rotary machine (Hoe, 1847).

Production speeds increased dramatically, and the volume of printed material grew to support the emergence of industrial society in countries around the world. As perhaps one of the earliest capitalist enterprises, the industrial development of printing from an individual craft to a manufacturing process was accompanied by a transition away from the relationships of the feudal hierarchy—the ranks of royalty, nobility, gentry, and other aristocratic titles—and toward the legal and business requirements of the rising entrepreneurial middle classes of the European towns and cities.

The moment of industrialization in the history of print technology is illustrated in the timeline by the separation of the left side (1400 to 1735) from the right side of the graphic (1735 to 2020) around the middle of the eighteenth century. It was during this transition from the Renaissance to the Enlightenment, from feudalism to capitalism, from preindustrial to industrial society that printing underwent a transformation such that the equipment, procedures, and roles played by the people in the process no longer resembled anything in Gutenberg's print shop in Mainz, Germany, save the letterpress method that transferred the ink onto the paper. The invention of the papermaking machine in a Paris suburb (Robert, 1798), during the French Revolution, contributed to this evolution and fed the consequent expansion of printed materials with more consistent substrates and improved quality.

The introduction of steam power (Koenig, 1803) during the second industrial revolution brought mass-manufacturing methods to printing with multilevel machines in large printing factories with teams of workers operating them. In the second half of the nineteenth century, the web press was introduced (Bullock, 1861), and the quantity of printing establishments in major cities, along with the number of daily newspapers, increased exponentially. The industrial expansion—especially in the US in the lead up to, during, and after the American Civil War—made fundamental improvements to the production of letterpress typography with the invention of the Linotype machine (Mergenthaler, 1886). The industrialization of printing took on the attributes of the assembly line methods associated with Henry Ford. The introduction of halftone photographic reproduction (Ives, 1883), the perfection of color printing, and the proliferation of typestyles were also driven by the needs of business advertising and the emergence of newspaper and magazine publishing empires to accommodate the thirst of the expanding urban populations for news and information.

Although electrification had been underway since the 1880s, the transition away from steam power did not fully take hold in factories and pressrooms until the second decade of the twentieth century. This electrification assisted the military needs of both World War I and World War II as significant developments were driven forward in communications technologies, and electronics were introduced into print machinery (Crosfield, 1947 and Hell, 1950) and brought with them methods of automation and remote controls. The postwar invention of the transistor (Shockley, 1947) and the subsequent revolution in digital technologies—first the integrated circuit (Noyce, 1959) and then microprocessors—transformed press controls as levers and knobs were replaced with PLCs (programmable logic controllers). Later, phototypesetting and the mechanical assembly of print content became transformed by proprietary computer electronic prepress systems (Arazi, 1979).

With the rise of the personal computer in the early 1980s, which enabled the transformative impact of desktop publishing (Jobs, 1985)—there will be more about this later—the entire printing process was revolutionized again with all the steps from the publisher to the pressroom replaced with software, lasers, and direct-to-plate imaging systems. With the invention of digital printing in the 1990s

(Landa, 1993), the computer transformation entered the pressroom, and the distinction between office equipment and industrial printing machinery became less defined. The growth and predominance of the Internet (ARPA, 1969), the World Wide Web (Berners-Lee, 1989), and wireless broadband connectivity in the early twenty-first century enabled the integration of printing with online order entry, proofing, and web-to-print technologies.

A similar analysis can be made of the development of the methodologies for imprinting an image onto a sheet of paper or other substrate over the past six centuries. The generally recognized printing methods—letterpress, intaglio or gravure, lithography, including offset lithography, screen printing, flexography, and xerography—correspond to different stages of socioeconomic development from craft manufacturing to industrial production, including the application of metal, rubber, and other synthetic materials that rely upon advanced chemistry, photomechanical, and electronic processes. The mechanical systems of letterpress (relief printing) and intaglio (recessed printing) are by far the longest surviving methods, each lasting for more than five centuries and well into the twentieth century. Although they continue to be used today for specialty printing purposes, they were essentially displaced by offset lithography (Senefelder, 1796 and Rubel, 1904) by the 1950s. The other three methods (i.e., screen printing [Simon, 1907] and flexography [Bibby, Baron and Sons, 1890] at the end of the nineteenth century and xerography [Carlson, 1950] in the middle of the twentieth century) were associated with the photomechanical and modern electronic processes, including the expansion of printing into areas such as the corporate and legal office, merchandising, retailing, and direct mail communications.

Ink-jet, the most recently developed method of placing a printed image onto paper (Howard, 1982), requires no image carrier or any intermediary transfer mechanism at all. Ink-jet is unique and heralds a new era of print technology because it places an image onto paper, or nearly any surface, directly with an array of nozzles or spray heads. The implications of this transition for the future of print are significant as it is a distinct departure from the static image reproduction of all previous methodologies, apart from xerography, and brings printing an infinitely variable image to a nearly limitless list of products and substrates. The integration of this digital process with big data repositories means that the mass manufacturing of any item can be imprinted with messaging that is relevant to small groups or subgroups of people or even a single individual.

As for the history of typographic design—a subject that is covered in many chapters of this book—it can be established that the style evolution over the centuries represents a complex interaction of both cultural and technical influences with the machinery, forms of print communications, and demographic changes. The first black letter fonts designed by Gutenberg and his contemporaries for book production during the incunabula (approximately 1450–1500) reflected the influence of the handwriting of the scribes and the matrix production limitations of the new metal-mold manufacturing technique. Connected with the artisanry of the individual printing establishment, typography underwent creative changes—roman (Jenson, 1470) and italic types (Manutius, 1501) and upper and lowercase characters being designed in the later decades of Gutenberg's lifetime—and then after the global expansion of printing, one form that the progression of the press took was the development of national fonts, such as Garamond (Garamond, 1495) in France. Later, the multiplicity of typeface styles and font families during the industrial era both expressed the various commercial and mass communication needs for signage, advertising (Goudy, 1903 and others), and the column inches of newspapers (Morison, 1930 and others). Although still essentially a letterpress era, the type design variety and volume of printed material exploited the speed and precision of the pantographic punch-cutting engraver (Benton, 1884) invented in the late nineteenth century.

With the introduction of phototypesetting (Moyroud and Higonnet, 1953) alongside of offset lithography in the twentieth century, type design underwent another transformation that corresponded with the esthetics of modernism. While serif typestyles still predominated in text-intensive book and publication design, sans serif types—such as Helvetica (Hoffmann, 1957), Univers (Frutiger, 1957), and Avant Garde Gothic (Lubalin, 1970)—proliferated after

World War II and dominated corporate identity, display type, and public information signage wherever the Latin alphabet and the English language were the international standard, which by this point was most of the world.

From this review of graphic communications history, it is seen that a logical evolution of stages exists in the development of printing-press machinery, methods of printing, image-transfer processes, phases of type production, and pictorial reproduction techniques. These have all proceeded along a path through individual inventors but also independently of them. That is to say, technological advancement in graphic communications has moved through periods of human history, and the individual innovators were "selected" such that the necessary breakthroughs were made at any given moment along this continuum. The specific identity of the innovators and how they were located is the product of a complex set of predetermined and accidental circumstances that found the right person at the right time.

This does not mean the determined vision of the inventors or the specific knowledge, skills, and talents they possessed did not play a role in this selection process. These subjective factors are extremely important and often become the final decisive element in causing the achievement to happen. However, in the end, these individual traits are themselves part of a concatenation of conditions, such as geographic location, social and economic environment, professional relationships, and the availability of resources that form the foundational background and driving forces of innovation.

The primary influence of socioeconomic background and geographic location on the seemingly random selection of individuals in the history of print technology can be confirmed, for example, by examining the transition of the center of worldwide innovation from Europe to America during the period of industrialization in the nineteenth century. Two key print technology advancements—the mounting of typeforms on a curved surface in the first rotary press (Hoe, 1847) and the successful implementation of web press cutoff technology (Bullock, 1861)—vastly increased the speed and output of powered printing presses. Both these developments, which

solved key barriers to industrialization that were known throughout the press equipment business, took place in the United States,
the former in New York City and latter in Pittsburgh, Pennsylvania.
While the industrial breakthroughs at the beginning of the 1800s
occurred in England (Stanhope, 1800) and Germany (Keonig,
1818), by mid-century, the role of American inventors in the conversion of pressrooms into mass production facilities with gigantic
presses operated by large crews of factory workers became dominant.
Nearly every hurdle in the way of the complete industrialization
of the printing process—the development of the rotogravure press
(Klíc, 1895) in Vienna, Austria, notwithstanding—took place in the
US, and this preeminence extended well into the twentieth century.

The force of innovation in the US was so great that even the
critical invention of the Linotype machine (Mergenthaler, 1886),
which automated the handwork of type compositors, took place
in Baltimore, Maryland. It is almost certain that had Ottmar
Mergenthaler not emigrated from Germany to the US at age of
eighteen in 1872, he would not have become the inventor of the
device known as the "eighth wonder of the world," and it stands
to reason that someone else would have solved this problem. The
changing dynamics of the world economy that accompanied the rise
of American industry were both anticipated and driven forward by
the revolution in print communications that exponentially increased
the volume and frequency of published materials needed to meet
the growth of urban populations, especially after the American Civil
War. In that era, the home of the great print publishing empires in
the industrial American Midwest and Northeast emerged as centers
of communication innovation on a par with what Silicon Valley has
become for electronic media a century later.

Senefelder made a highly significant comment in his autobiographical account when he explained that had he possessed the
financial resources to invest in "types, a press, and paper," then lithography "probably would not have been invented so soon." Senefelder
was acknowledging that the invention of the new purely chemical
printing method was historically inevitable and may well have been
made by someone else had it not been for the random circumstance

of his inability to purchase the resources required to start a letterpress printing business.

Senefelder understood that it was a matter of time before the discovery of lithography would have happened anyway. After all, his invention coincided with major advancements in chemistry in the late eighteenth century. Known as the "chemical revolution," the elements of water and air molecules were being defined, and atomic theory was being suggested while Senefelder was experimenting with the interaction of various chemicals and printing-plate surfaces with the use of an oil-based writing tool.

It is interesting to note that, while popular accounts of Senefelder's invention of lithography often make mention of the "grease pencil" or a "greasy crayon" he used to write the laundry list upon the limestone, there is hardly a reference to the nature of this instrument or the substances out of which it was made. Readers of the stories of the accidental invention of lithography by Senefelder will not find an explanation of the significance his possession of a crayon while he had a slab of limestone even though he mentioned his use of this writing tool throughout his biographical account.

Senefelder described the various compositions of the crayon that he experimented with for the purpose of applying it "on the stone plate in dry form like Spanish or Parisian chalk." He gave eight different combinations of wax, soap, lampblack, spermaceti (whale oil), shellac, and tallow from which he created the prepress writing instrument. He wrote that, "Wax, mennig, and lampblack are heated and constantly stirred till the mennig dissolves in froth and changes from red to brown. Then the lampblack is rubbed in thoroughly, the whole warmed again properly, and shaped into sticks." This is one of the earliest descriptions of crayon use for artistic purposes, and to some extent, it can be said that Senefelder is both the inventor of lithography and the crayon.

In a similar manner, while there are repeated references to the misfeed jam of Rubel's lithographic press that had a "rubber blanket" on its impression cylinder, little time has been devoted to a study of how it came to be that his press contained such a blanket system. The rubber on Rubel's press was most certainly of the natural and vulcanized type that had been developed and perfected by Charles Goodyear (US) and Thomas Hancock (UK) by the middle of the previous century. While products, such as printing cylinder blankets, were derived from rubber trees grown in tropical and sub-tropical regions of South America, Western Africa, and Southeast Asia, the material tended to swell and blister under the pressure and heat of the printing process. Synthetic rubbers, especially heat-resistant neoprene, were invented in the 1930s and solved many of these issues. Without the circumstances that brought rubber blankets to lithographic press machinery and the subsequent improvement and perfection of these blankets with synthetic materials that took place before and during World War II, the offset method would not have been invented at the turn of the century, and the rise of offset lithography would have been delayed.

From Johannes Gutenberg to Steve Jobs

While reviewing nearly six centuries of print technology—through the lives and inventions of significant industry innovators—it became clear that the invention of printing by Johannes Gutenberg on the one side and the breakthrough of desktop publishing by Steve Jobs on the other are bookends in the age of print. While it has long been acknowledged that the handheld-type mold and printing press are the alpha in the age of manufacturing production of ink-on-paper forms, such as books, newspapers, magazines, etc., the view that desktop publishing is the omega of this age is not necessarily widely held. When viewed within the framework of disruptive continuity, it can be shown that the innovations of Gutenberg and Jobs manifest similar attributes in terms of their dramatic departure from previous methods as well as their connection to the multilayered processes of cultural changes in the whole of society in the fifteenth and twentieth centuries.

The advent in 1985 of desktop publishing—a term coined by the co-founder of Aldus Corporation, Paul Brainerd—is associated with Steven P. Jobs because he contributed to its conceptualization, he articulated its historical significance, and he was the innovator who made it into a reality. With the support of publishing-industry consultant John Seybold, Jobs went on to integrate the technologies and brought together the people that represented the elements of desktop publishing: a personal computer (Apple Macintosh), page-layout software (Aldus PageMaker), page-description language (Adobe PostScript), and a digital laser printing engine (Canon LBP-CX). He demonstrated the integration of these technologies to the world for the first time at the Apple Computer annual stockholders meeting on January 23, 1985, in Cupertino, California, a truly historic event in the development of graphic communications.

It is a fact that the basic components of desktop publishing had already been developed in the laboratory at the Xerox Palo Alto Research Center (PARC) by the late 1970s. However, due to a series of issues related to timing, cost, and the corporate culture at Xerox, the remarkable achievements at PARC—which Steve Jobs had seen during a visit to the lab in 1979 and inspired his subsequent development of the Macintosh computer in 1984—never saw the light of day as commercial products. As is often the case in the history of technology, one innovator may be the first to theorize about a breakthrough, or even build a prototype, but never fully develop it while another innovator creates a practical and functioning product based on a similar concept and this becomes the wave of the future. This was certainly the case with desktop publishing where many of the elements that Jobs would later integrate at the Cupertino demo in 1985—the graphical user interface (GUI), the laser printer, desktop software integrating graphics and text, what-you-see-is-what-you-get (WYSIWYG) printing—were functioning in experimental form at PARC at least six years earlier.

What became known as the desktop publishing revolution was just that. It was a transformative departure from the previous photomechanical stage of printing technology on a par with the separation of Gutenberg's invention of mechanized metal-printing type production from the handwork of scribes. Desktop publishing brought the era of phototypesetting that began in the 1950s to a close. It also eventually displaced the proprietary computerized prepress systems that had emerged and were associated with Scitex in the late 1970s.

Furthermore, and just as significant, desktop publishing pushed the assembly of text information and graphic content beyond the limits of ink-on-paper and into the realm of electronic and digital media. Thus, desktop publishing accomplished several things simultaneously: (1) it accelerated the process of producing print media by integrating content creation—including the design of text and graphics into a single electronic document on a personal computer—with press-manufacturing processes; (2) it expanded the democratization of print by enabling anyone with a personal computer and laser printer to mass produce and distribute printed material starting with a copy of one; (3) it created the basis for the personalization of print media; and (4) it laid the foundation for the expansion of a multiplicity of digital media forms within a decade, including electronic publishing in the form of the Portable Document Format (PDF), e-books, interactive media, and ultimately, contributed the global expansion and domination of the Internet and the World Wide Web.

Just as Gutenberg attempted to replicate in mechanized form the handwriting of the scriptoria, the initial transition to digital and electronic media by desktop publishing carried over the various formats of print—that is, books, magazines, newspaper, journals, etc.—into digital files stored on magnetic and optical storage systems, such as computer magnetic and optical disks. However, the expansion of electronic media—which was no longer dimensionally restricted by page size or number of pages but limited by data storage capacity and the bandwidth of the processing and display systems—brought the phenomenon of hyperlinks and drove entirely new communications platforms for publishing text, graphics, photographs, and audio and video eventually on mobile wireless devices with content stored on cloud computing servers within just two decades of Jobs's 1985 innovation. Through websites, blogs, streaming content, and social media, suddenly every individual could record and share their life story, be a reporter and publisher, or participate in, comment on, and influence events in real time anywhere in the world.

One need only take note of the ten-minute smartphone video captured by seventeen-year-old Darnella Frazier on the street corner outside of Cup Foods grocery store in Minneapolis, Minnesota, on

May 25, 2020, to recognize the impact of these changes upon society. After she posted the video of the killing of George Floyd by former police officer Derek Chauvin on Facebook and Instagram at one forty-six the next morning, a mass movement developed involving as many as twenty million people in more than two thousand cities and towns in over sixty countries throughout the world. On June 11, 2021, Frazier was presented with a special award and citation from the Pulitzer Prize board for her courage and for "highlighting the crucial role of citizens in journalists' quest for truth and justice."

The groundbreaking significance of desktop publishing, which straddled both the previous printing and the new digital media ages, can be further illustrated by going through the above description by Will Durant of the impact Gutenberg's invention and substituting the new media for printing and the other contemporary elements of social, intellectual, and political life for those the historian identified in the 1950s about the fifteenth century:

> Soon more than half the world's population was both consuming and publishing their own content from smartphones and tablets as well as personal computers, which became one of the effervescent ingredients of the information age…the digital revolution was on.
>
> To describe all the effects of desktop publishing and electronic media would be to chronicle well more than half the history of the modern mind…. It replaced all informational print by republishing it online as text or in more complex graphical formats like PDF, with methods for managing versions and protecting authenticity so that scholars and researchers in diverse countries may work with one another through video streaming and virtual reality tools, allowing entry of new information and data to be gathered, published, and shared in real time as though they were sitting in the same room…. Online

electronic media made available to the public all the world's manuals and procedural instructions; with the development of international collaborative projects, such as Wikipedia, it became the greatest tool for learning that has ever existed at no charge and available to all. It did not produce Modernism or the information age, but it further paved the way for a new stage of human society that had been promised by the American and French Revolutions based on democracy and where genuine equality exists as a fundamental right for everyone. It made the entire library of literature, music, fine and industrial arts, architecture, theater, athletic competition, and cinema instantly available anywhere and at any time in the palm of the hand and prepared the people for an understanding of the role of mythology, mysticism, and superstition in history by demonstrating the application of a scientific and materialist outlook in everyday life. It ended the monopoly of news by corporate and state publishers and the control of learning by educational institutions managed by the prevailing ruling establishment. It encouraged the streaming of live video by anyone to the entire world's audience of mobile device owners that could never have been reached through printed media. It facilitated global communication and cooperation of scientists and enabled the launching of the International Space Station and the sending of multiple probes to the surface of Mars and beyond. It affected the quality and character of all published literature and information by subjecting authors and journalists to the purse and taste of billions of regular working people in both the advanced- and lesser-developed countries rather than to just

the middle and upper classes. And, after speech
and print, desktop publishing, online, and social
media provided a readier instrument for the dissemination of nonsense and disinformation than
the world has ever known.

Up to the present, the new media has not yet displaced print
the way print eventually replaced the scribes. Even when it does, it
is likely that printing on paper will continue to exist well into the
future in a similar manner in which the ancient art of pen-and-ink
calligraphy has continued to exist alongside of print for centuries
long after the last scriptorium was shut down. Meanwhile, electronic
media such as e-books have contributed to a resurgence of printed
books, and after the initial fascination with the electronic devices
such as Amazon's Kindle, the popularity and thirst of the public for
print has increased, particularly following the onset of the coronavi-
rus pandemic. While the forced separation of people from each other
has driven up the use of digital tools, such as online video meetings,
events, and gatherings, the self-isolation of reading a printed book
suddenly peaked again.

Considerable effort has been made to replicate the experience
of reading print media in electronic form. The advent of e-paper—
the simulation of the look and feel of ink-on-paper which was pioneered at Xerox PARC in the 1970s (Sheridon, 1974)—is an attempt
to adapt two-dimensional digital display technologies to mimic the
reading experience of the printed page. Since then, studies have
shown that paper-based books yield superior reading retention to
that of e-books. This is not so much because of the appearance of
the printed page and its impact on visual perception as it is the tactile experience and spatial awareness connected with turning physical
pages and navigating through a three-dimensional volume that contains a table of contents and an index.

In 2012, during a presentation at the DRUPA International
Printing and Paper Expo in Düsseldorf, Germany, Benny Landa, the
pioneer of digital printing who developed the Indigo Press in 1993,
said the following, "I bet there is not one person in this hall that

believes that two hundred years from now, man will communicate by smearing pigment onto crushed trees. The question on everyone's mind is when will printed media be replaced by digital media.... It will take many decades before printed media is replaced by whatever it will be…many decades is way over the horizon for us and our children."

Since Landa's talk at DRUPA was part of the introduction of a new press with a digital printing method called nanography, he was emphasizing that we need to live and work in the here and now and not get too far ahead of ourselves. Landa's nanographic press is based on advanced imaging technology that transfers a film of ink pigment to almost any printing surface which is multiple magnitudes thinner than either offset or other digital presses. By removing the water in the ink-jet process, the fusing of toner to paper in xerography and the petroleum-based vehicles that carry pigment in traditional offset presses, nanography dramatically reduces the cost of reproduction by focusing on the transfer of ultrafine droplets of pure pigment (nanoink) first to a blanket and then to the substrate. The aim of nanography is to keep paper-based media economically viable by providing a variable imaging press that can compete with the costs of offset lithography and accommodate the needs of the hybrid digital and analog commercial printing marketplace.

While global print advertising markets are in decline, society is not yet ready to make a full transition to electronic media and move entirely away from paper communications. This is a serious dilemma facing those working in the printing industry who are trying to navigate the difficulties of maintaining a viable business in an environment where print remains in demand—in some segments it is growing—but overall, it is a shrinking percentage of economic activity. With greater numbers of people and resources being redirected to communications and marketing products in the more promising and profitable big tech and social media sectors, the printing industry is being starved of talent and economic resources.

Rather than trying to put a date on the moment of transition to a post-printing and fully digital age of communications, the more relevant question is how it will be accomplished. Landa had it right when he said that today most people believe that two hundred years from now, man will no longer communicate by "smearing pigment onto crushed trees." When the character of print media is put in these terms, the historical distance of the present analog form of communications from the long-term potential of the digital age becomes clearer. Still, no vision or road map has yet been articulated for what is required for civilization to elevate itself beyond the age of print.

It is difficult to discuss the moment of a complete progression of human communication methods from Gutenberg to Jobs without reference to the work of the Canadian media theorist Marshall McLuhan. Although McLuhan's presentation lacked a coherent perspective and tended to drift about in what he called the "mosaic approach," he made numerous prescient observations about the forms of media and the evolution of communications technology. Sharing elements of the theory of disruptive continuity, McLuhan focused in on the reciprocal interaction of the modes of communication—spoken, printed, and electronic—with the broader societal, economic, cultural, and ideological transformations in world history. He emphasized the way these transitions each fundamentally altered man's consciousness and self-image. He also recognized that there was presently a "clash" between what he called the culture of the "electric age" with that of the age of print.

During an interview with the British Broadcasting Corporation in 1965, McLuhan explained how he saw technology as an extension of man's natural capabilities:

> If the wheel is an extension of feet, and tools of hands and arms, then electromagnetism seems to be in its technological manifestations an extension of our nerves and becomes mainly an information system. It is above all a feedback or looped system. But the peculiarity, you see, after the age of the wheel, you suddenly encounter the age of the circuit. The wheel pushed to an extreme suddenly acquires opposite characteristics. This seems to happen with a good many

technologies—that if they get pushed to a very distant point, they reverse their characteristics.

This is an early reference by McLuhan to the omnidirectional character of electronic media. While print media is associated with the acceleration of visual culture and the creation of national entities extending far beyond the influence of handwriting by the scribes, it displaced auditory-based spoken-word culture and also brought into existence a unidirectional or one-to-many communications platform of information and knowledge. The press was powerful but ensconced within the limitations of the socio-economics of the national state framework. Electronic media contains within it all of the progressive elements of print and restores the power of the spoken word alongside visual communications with a multiplicity of one-to-one, one-to-many, and many-to-many channels that spans the globe and instantaneously penetrates the limits of all previous national boundaries.

Among McLuhan's most significant contributions are found in his 1962 work, *The Gutenberg Galaxy: The Making of Typographical Man*. He discusses the reliance of primitive oral culture upon auditory perception and the elevation of vision above hearing in the culture of print. He wrote that his study "is intended to trace the ways in which the forms of experience and of mental outlook and expression have been modified, first by the phonetic alphabet and then by printing." For McLuhan, the transformations from spoken-word culture to typography and from typography to the electronic age extended beyond the mental organization of experience.

In the preface to *The Gutenberg Galaxy*, McLuhan summarized how he saw the interactive relationship of media forms with the whole social environment:

> Any technology tends to create a new human environment. Script and papyrus created the social environment we think of in connection with the empires if the ancient world.... Technological environments are not merely passive containers

of people but are active processes that reshape people and other technologies alike. In our time, the sudden shift from the mechanical technology of the wheel to the technology of electric circuitry represents one of the major shifts of all historical time. Printing from movable types created a quite unexpected new environment—it created the public. Manuscript technology did not have the intensity or power of extension necessary to create publics on a national scale. What we have called "nations" in recent centuries did not, and could not, precede the advent of Gutenberg technology any more than they can survive the advent of electric circuitry with its power of totally involving all people in all other people.

As early as 1962—seven years before the creation of the Internet and nearly three decades before the birth of the World Wide Web—McLuhan anticipated the historical, far-reaching, and revolutionary implications of the information and electronic age on the global organization of society. Although he eschewed determinism in any form, McLuhan pointed to the potential for electronic media to drive mankind beyond the national particularism which is rooted in the technical, sociohistorical, and scientific eras connected with the age of print. McLuhan later used the phrase "global village" to describe his vision of a higher form of nonnational organization driven by the methods of human interaction that were brought on by "the advent of electric circuitry" and "totally involve all people in all other people." For McLuhan, the transformation from the typographic and mechanical age to the electric age began with the telegraph in the 1830s. The new media created by the properties of electricity were expanded considerably with telephone, radio, television, and the computer in the nineteenth and twentieth centuries. McLuhan also wrote that the electronic media transformation revived oral culture and displaced the individualism and fragmentation of print culture with a "collective identity."

McLuhan's examination of the historical clash of the electronic media with the social environment of print culture and his prediction that a new collective human identity will be established from the transition to a global structure beyond the present fragmented national identities is highly significant. It points to the coming of the transformations that will be required for electronic media to thoroughly overcome print media as a completed historical process. In a similar way that Gutenberg's invention spread across Europe and the world and planted the seeds of foundational transformation—in technology, politics, and science—that developed over the next three and a half centuries, we are today, likewise, in the incubator of the new global transformation of electronic media.

With this historically dynamic way of understanding the present, the worldwide spread of smartphones and social media to billions of people, despite national barriers placed upon the exchange of information as well as other differences, such as language and ethnicity, humanity is being transformed with the emergence of a new homogeneous global culture. For this development to achieve its full potential, the social organization of man must be brought into alignment, and there is no reason to believe that this adjustment from nations to a higher form of organization will take place with any less discontinuity than that of the period of world history that began with the rapid development of printing technology and concluded with the American and French Revolutions.

There are scientists and futurists today who either proselytize or warn about the coming of the technological singularity—that is, the moment in history when electronic media convergence and artificial intelligence will completely overtake the native capacities of humanity. The argument goes that these extensions of man will become irreversible, and civilization will be transformed in unanticipated ways either toward a utopian or dystopian future, depending on whether one supports or opposes the promises of the singularity. The twentieth-century philosophical and intellectual movement known as transhumanism promotes the idea that the human condition will be dramatically improved through advanced artificial intelligence tools, robotics, and cognitive enhancements. The dystopian opponents of transhumanist utopianism argue that technological advancements, such as artificial intelligence, should not be permitted to supplant the natural powers of the human mind on the grounds that they are morally compromising and pose an existential threat to society. Among these competing views, however, is the shared conception that the coming transformation of mankind will take place without a fundamental change in the socio-economic environment. Both the supporters and opponents of transhumanism envision that the extensions of man will evolve independently of any realignment of the foundations of society.

Keeping these essential limitations of both the advocates and opponents of transhumanism in mind, it is not possible to prognosticate about the future of communications media independently of an understanding that the impulses for broad and fundamental societal change that were embedded in the technology of print nearly six centuries ago—especially the democratization of information and knowledge—have been vastly expanded and taken to a qualitatively new level in our own time. In a world where every individual has the potential to communicate as both publisher and consumer of information with everyone else on the planet—regardless of geographic or national location—it appears entirely possible and necessary that new and higher forms of social organization must be achieved before this new media can carve a path to a truly post-printing age of society. While the existential threats are real, they do not come from the technology itself. The danger arises from the clash of the emerging global social structures and self-identity of mankind—driven, in part, by the electronic and instantaneous online, mobile, and social media communications landscape—against the national socioeconomic and state institutions and the ideological influences associated with the previous era of print communications.

1

Johannes Gutenberg
c. 1398–1468

Handheld Mold for Metal Typecasting
Mainz, Germany
1440

T he known details of Johannes Gutenberg's life are few and far between. With documentary evidence quite meager after nearly six centuries, there are many gaps in Gutenberg's biography. Due to the lack of information, a mythology has been built up around Gutenberg that (1) his ideas about printing came to him "like a ray of light," (2) that he was a failed businessman, and (3) he died in poverty. None of these facts are true.

What is known is that Johannes Gutenberg left Mainz in 1430 due to political conflicts between the patricians and the guilds. Gutenberg, himself a patrician with an inclination toward the guild members, was owed considerable sums by the local government. It is likely that he began his project in 1439 while living in Strasbourg. Far from it coming to him in an instant, Gutenberg worked on what he called his "secret enterprise" for some ten years before it was complete and ready for commercial production. The processes contained in his innovation were complex and expensive and would have required numerous approaches and attempts. Among them were (1) typeface design, (2) engraving of patrices, (3) manufacture of matrices, (4) creation of the manual metal typecasting mold, (5) composition of metal alloys, (6) ink formulation, (7) experiments with paper and parchment, and (8) the construction of the wooden press machinery.

By far, the most significant of these was (4) the invention of the handheld mold for casting metal type. There are recreations of the invention at the Gutenberg Museum in Mainz, Germany, but none of the original casting devices have been preserved; the ones in the museum were rebuilt from information available about how they were constructed. This information did not include any drawings or schematics.

It is believed that Gutenberg returned to Mainz in 1448, and it was around this time that the process was finalized, and live projects could be produced with his invention. In 1449–50, Gutenberg secured an investment from Johann Fust, and the two became partners, opening the first commercial printing establishment in the world. A rented facility was located, new presses were built, a staff was hired and trained, materials were procured and stored for the purpose of producing the forty-two-line Bibles that are well-known.

In 1455, there was a business dispute between the two men, and Fust sued Gutenberg in court on charges of refusal to pay interest on his loan and embezzlement. In a complex ruling, the court issued an order for Gutenberg to pay a portion of what Fust demanded, and the two parted company. The legal dispute with Fust certainly set Gutenberg back as he was unable to pay immediately. Fust kept the Bible inventory, opened his own printing facility, and took the most skilled employee of the firm, Peter Schöffer, with him. However, Gutenberg was not ruined, and he continued to work energetically on the development of his technique…he just had a competitor down the road, another first in the industry.

It is believed that Gutenberg continued to produce Bibles and other products, such as calendars and letters of indulgence. In 1465, the archbishop of Mainz, Adolf von Nassau, appointed Gutenberg as "gentleman of the court" in recognition for his achievements. Gutenberg enjoyed the privileges of the recognition until his death in 1468. His invention spread rapidly throughout Europe, led to the expansion of literacy, and is considered one of the greatest achievements of the Renaissance.

2

Nicolas Jenson

c. 1420–1480

Roman Type
Venice, Italy
1470

The term *incunabula* (Latin for "cradle") is used to denote the earliest period of printing from its birth in 1450 up to January 1, 1501. The books, pamphlets, and broadsides printed with the movable metal type method associated with Gutenberg during these first fifty years are also commonly called incunabulum.

It is estimated that thirty-five thousand editions were printed throughout Europe—over two-thirds from Germany and Italy—during the second half of the fifteenth century. Remarkably, nearly 80 percent of these volumes still exist today, most of which are held in large public collections, such as the Bavarian State Library in Munich, the Vatican Library in Vatican City, and the British Library in London.

The most famous incunabulum, of course, is the forty-two-line Bible printed by Johannes Gutenberg in Mainz, Germany, in the 1450s of which there are forty-eight copies remaining. Since they were printed in two volumes, many of these copies are incomplete. James Lenox brought the first complete set of the *Gutenberg Bible* to the United States in 1847 after he bought it for $2,500; it now sits on display at the New York Public Library. The last sale of a complete

Gutenberg Bible took place in 1978 and went for $2.2 million; it is estimated that one would sell for $25–$35 million today.

The British Library maintains an international electronic bibliographic database of extant incunabulum. Called the Incunabula Short Title Catalogue (ISTC), the database was begun in 1980 and currently contains 27,460 records. The ISTC is an extraordinary merger of modern and Renaissance information technology. That anyone can peruse these records—many of which have links to high-resolution images of five-hundred-year-old incunabulum—is a testament to both the lasting achievement of print and the significance of its electronic descendent, the World Wide Web.

*****

Next to Gutenberg himself, Nicolas Jenson is recognized as the most important figure of the incunabula. Despite limited records of his life—his last will and testament, a few book introductions written by others, and some document fragments—the legacy of Nicolas Jenson survives through his printed works.

According to Martin Lowry, the printing scholar and author of *Nicholas Jenson and the Rise of Venetian Publishing in Renaissance Europe*, the first official biography of Jenson was written in the late 1700s and amounted to "a two-volume potpourri of erudition and fantasy." While arguing that Nicolas Jenson has become something of a printing cult-figure, Lowry does conclude that the Jenson's "place at the very beginning of the typographic age gives him a special importance."

It is known that Nicolas Jenson was born in Sommevoire, France, a town about 150 miles southeast of Paris. However, after reviewing Lowry's research, it is difficult to simply repeat here the many other "facts" that are frequently given of Jenson's early life: his date of birth, his employment experience, the origin of his metal-working skills, how he became familiar with the printing methods of Gutenberg, and his route from France to Italy. The things that are repeated in many accounts of Jenson's life are derived from murky historical anecdotes that are contradicted by other important facts.

Nicolas Jenson is known to have begun printing in Venice in the late 1460s or early 1470s. Prior to his arrival in Venice, it appears that he spent some time in Vicenza, a mainland town about thirty miles to the west, where he developed his printing skills. Jenson's arrival in Venice, the first non-German printer in recorded history, coincided with the establishment of several important printing firms in the Italian island city. The most notable of these was the enterprise of the John and Wendelin of Speyer who arrived in Venice from Germany in 1468 and were granted a five-year monopoly on printing by the city authorities.

The Venetian patrician class of scholar-statesmen considered the arrival of printing a major cultural development. It meant that the works of classical humanist teachings could be reproduced at rates that were inconceivable with the handwritten process of the scribes. The ruling elites encouraged the development of print, and by the end of the century, there were 150 firms operating in the highly competitive Venetian printing market.

Alongside of print's cultural impact, there was a considerable business opportunity to be exploited. It is to this side of the incunabula that Jenson devoted most of his efforts. During the ten years that he was a printer in Venice, more than anyone else, Jenson brought investment into the printing industry. His businesses were very successful, and he made a considerable fortune before his death in 1480.

However, the most important, and universally recognized, contribution of Nicolas Jenson to the development of printing was his design of an early roman typeface. Prior to Jenson, the style of print typography followed the black letter example set by Gutenberg—that is, heavy gothic forms that emulated the dominant pen and ink script of the monks of fifteenth century Germany.

Such were Nicolas Jenson's metal-working skills that he cut a groundbreaking roman type in 1470. Roman type is distinct from black letter in that it emulates the square capital letters used in ancient Rome combined with the Carolingian minuscule (lowercase) used during the Holy Roman Empire. The first book to appear with Jenson's new design was an edition of Eusebius's *Preparation for the Gospel* originally written in AD 313.

The word roman, without a capital R, has come to denote Italian typefaces used during the Renaissance as well as later fonts derived from them, such as Times New Roman, for example. Although Jenson's design was quite different in appearance from Gutenberg's black letter, it was also modeled on the scribal manuscript style that was popular in fifteenth century Italy.

It is a remarkable phenomenon of printing history that the essential forms of Jenson's roman typeface designed more than five hundred years ago are those that we continue to use most often and recognize today as the best and most readable typography. Of course, the characters in the alphabet of the Latin languages are those associated with Jenson's contribution. But it should also be noted that Jenson designed and cut a Greek alphabet of a similar style.

Throughout the subsequent history of printing, many have noted the beauty and balance of Jenson's roman-type design. William Morris and the arts and crafts movement of the late nineteenth century focused upon Jenson's creative genius. According to Lowry, Morris's romantic affinity for medievalism led to an unjustified elevation of the contribution of Nicolas Jenson alongside those of Johannes Gutenberg and Aldus Manutius.

*****

A search of the British Library's ISTC for the term "Jenson" results in 113 hits. Many of the items in the database contain links to images of the pages printed by Nicolas Jenson himself on a Gutenberg-style printing press in Venice in the 1470s. A review of these entries shows that, despite language challenges, Jenson's books appear very similar to those found today in our libraries and bookstores. While some of them are adorned with ornate case-bound covers and others include hand-illuminated art alongside the printed text, the essential elements of the book are very familiar to any reader today.

Historians have strictly defined the incunabula as the first fifty years of the printing revolution beginning with Gutenberg. The incunabulum produced by the pioneers of print, including Nicolas Jenson, were devoted to a recreation of scribes' handwriting such that the reading audience could understand and relate to the new media form.

The questions that arise naturally are should we consider the early years of the digital revolution to be our modern "incunabula" in which the previous media generation is being replicated in electronic form, or is the digital age leading to a new media that represents a departure from the forms that were developed and enriched during the Renaissance? It will take some time for these questions to be answered definitively.

3

Aldus Manutius
1449/1452–1515

Italics and Octavo Book Format
Venice, Italy
1501

The great cultural movement called the Renaissance (rebirth) spanned the fourteenth to the seventeenth centuries. This places Johannes Gutenberg's 1450 invention of printing early in that era. As is widely acknowledged by historians, printing was not only the most important technological achievement of the Renaissance, but it was also its greatest catalyst.

Along with the expansion of world trade and the circumnavigation of the globe—combined with enthusiasm for Gutenberg's innovation—printing spread rapidly from Germany to destinations across Europe and elsewhere. And as the business and products of "mechanical writing" proliferated, the knowledge that came along with it multiplied exponentially.

Eventually, like so many printing-press pebbles tossed into the global pond, the ripple waves of literacy and democracy spread and connected with one another. By the eighteenths century, the achievements of the Renaissance led to the age of Enlightenment and one of the most remarkable end results of this technical and cultural transformation was the American Revolution of 1776.

The Renaissance, especially in Italy, was shaped by humanist teachings—that is, the notion that citizens should be educated in the humanities (language, literature, philosophy, religion, and the arts). Florence, Naples, Rome, Venice, and Genoa were among the centers of Italian humanism. Great collections of antique handwritten manuscripts and printed books were assembled in libraries and made available for study primarily for those of wealth and means.

It was within this environment that Aldo Manuzio (Aldus Manutius is Latinized) was born in Bassiano, Italy, about one hundred kilometers south of Rome. The precise date of his birth is not known. It has been surmised that he was born sometime between 1449 and 1452 from the preface to a book published by Aldus's grandson, Aldus the Younger, in 1597.

Very little is known about Aldus's early life. In his *The Rudiments of Latin Grammar* published in 1501, Aldus mentions he was trained in Latin at a young age. This information—along with the fact that he had an ancestor that was a bishop—indicates that his family was aristocratic.

Aldus left Bassiano for Rome, perhaps as young as fifteen years of age, to be educated as a humanist scholar. His classical studies in Latin continued in Rome for eight years at which time he moved to Ferrara to conduct studies in Greek. Sometime around 1479 or 1480, having established himself as a scholar of the highest quality, the king of Carpi hired Aldus to become the teacher of his nephews. It was during his time in Carpi that Aldus developed his interest in publishing and printing.

By the 1400s, Venice was a center of world commerce much like New York City might be considered today; everything important was happening there. And so it was to this Italian center of so many things that Aldus, at the time in his late thirties, decided to relocate and start a publishing enterprise.

There is almost no information available about Aldus's activity during his first five years in Venice. It has been deduced that he spent this time preparing and setting up a viable publishing business—learning the book market and printing technique—in a very competitive environment. Printing arrived in Venice by way of Germany twenty years earlier, and there were well-known firms already operating by the time Aldus launched his enterprise.

Aldus's initial activity—and this would prove to be his most important accomplishment—was to edit and republish authoritative editions of the classics of literature. The initial project, a reissue of the *Greek Grammar* of Constantine Lascaris in 1495, was the first volume to be published under Aldus's name although Andrea Torresani's press likely did the printing.

A big breakthrough came in that same year with Aldus's publication in Greek of the first of a five-volume folio edition of the works of Aristotle. It would take four years to complete the project. This work has been referred to as the "greatest scholarly and printing achievement of the fifteenth century." Aldus then went on to edit and publish Thucydides, Sophocles, and Herodotus in 1502, Xenophon's *Hellenica* and *Euripides* in 1503, and Demosthenes in 1504. In the decade before his death, Aldus also published editions of Latin and Italian classics.

By 1496, Aldus had his own printing operation and began using various forms of "at Aldo's" to signify the source of his publications and what later became known as the Aldine Press. In 1498, Aldus began using the Aldine Press trademark—the emblem of the dolphin wrapped around the anchor that symbolized the Latin phrase "*Festina lente*" (Make haste slowly)—on all his works. At this point, Aldus turned his attention to innovations in the forms of print, the first big contribution being in typography.

The typography of print as it arrived from Germany was in the gothic form of heavy, "gross" lettering as seen in the *Gutenberg Bible*. By 1470, Nicolas Jenson, who had come to Venice from France, had transformed the printed word and created the Renaissance book with his slender roman typeface and capital letters.

Aldus began with Jenson's typeface and reworked it into his own style for the Greek classics. His desire was to make the fonts look like the work of human handwriting. This effort would ultimately lead to the development of the "italics" that emulated cursive writing and became known at times as the Aldino typeface. Francesco Griffo, to whom Aldus later paid tribute, performed the actual artistry of cutting the slanted italic. The first book that contained Aldus's italic font was the 1501 Aldine Press edition of Virgil's *Opera*.

During the incunabula era, from 1450 to 1500, most books were printed in folio format—that is, page sizes of approximately 14.5" × 20" (the pages in the *Gutenberg Bible* are approximately 12" × 17.5"). These were very large books that could be read wherever they were located; they were not portable.

This is when Aldus made another historic advancement. As explained by Helen Barolini in her *Aldus and His Dream Book*, "The real revolution, the moment of true divulgation of the printed word that impelled Western society…came when Aldus brought out his edition of Virgil's *Georgics* in elegant, octavo format that was to become the staple of the Aldine Press and Aldus's trademark."

The octavo format is approximately 7.25" × 10", something very close to what would be found today in a bookstore. Aldus produced the very first modern book that was small enough and inexpensive enough for someone to take with them almost anywhere.

Aldus's other innovations included a unique method for bookbinding, sometimes referred to as "binding in the Greek style," and advancements in punctuation, such as the creation of the semicolon and the modern comma.

Aldus Manutius married Maria, the daughter of Andrea Torresani, the owner of a printing firm with whom he had worked early on in Venice, in 1505. Aldus died on February 6, 1515. His brothers-in-law took over the Aldine Press and ran it until 1533 when Aldus's third son, Paulus Manutius, assumed control of the business. It is believed that the Aldine Press published more than one thousand titles in the hundred years that ended in 1595.

The scholar, humanist, publisher, and printer Aldus Manutius was a monumental figure of the Renaissance. He was a force for the expansion of literacy and knowledge for everyone. Fortunately, his books have survived, and there is a substantial record of his work including his own words that were often printed in the preface of Aldine Press books.

In the Thesaurus Cornucopiae of 1496, Aldus wrote, "My only consolation is the assurance that my labors are helpful to all…so that even the 'book-buriers' are now bringing their books out of their cellars and offering them for sale. This is just what I predicted years ago when I was not able to get a single copy from anyone on loan, not even for one hour. Now I have got what I wanted: Greek volumes are made available to me from many sources… I do hope that, if there should be people of such spirit that they are against the sharing of literature as a common good, they may either burst of envy, become worn out in wretchedness, or hang themselves."

4

————

Benjamin Franklin
1706–1790

The Public Library
Philadelphia, Pennsylvania
1731

At the time of his death in 1790, Benjamin Franklin was world-famous as a philosopher, scientist, inventor, and diplomat. His significant contributions to the American Enlightenment and as a leading figure of the Revolution of 1776 are far too numerous and important to be appropriately dealt with in this short space. However, his work as a printer deserves special appreciation.

In 1728, when he was twenty-two years old, Benjamin Franklin wrote the following humorous epitaph for himself:

> *The body of*
> *B. Franklin, Printer*
> *(Like the Cover of an Old Book*
> *Its Contents torn Out*
> *And Stript of its Lettering and Gilding)*
> *Lies Here, Food for Worms.*
> *But the Work shall not be Lost;*
> *For it will (as he Believ'd) Appear once More*
> *In a New and More Elegant Edition*

Revised and Corrected
By the Author.

In his last will and testament of 1788, Franklin amended his plan for the inscription on his gravestone to bear just the names of himself and his wife, Deborah Read Franklin, who preceded him in death by fifteen years. In his will, however, he identified himself thus: "I, Benjamin Franklin, of Philadelphia, printer…"

Born in Boston, Massachusetts, Benjamin Franklin was the last of his father's seventeen children. Franklin began his printing career when he was apprenticed to his brother, James Franklin, at age twelve. After he was denied his wish to be published in James's newspaper, the young Ben began submitting letters to *The New-England Courant* under the pseudonym "Mrs. Silence Dogood." The correspondence became very popular in the local community.

At age seventeen, Benjamin left his apprenticeship without permission and ran away to Philadelphia to start out on his own. After building his reputation as a skilled craftsman, as both a type compositor and pressmen, and disciplined worker, Franklin had the opportunity to set up a printing house in partnership with Hugh Meredith in 1728. The following year, he became publisher of *The Pennsylvania Gazette*, one of two newspapers in the colonies.

In his Philadelphia printing shop, Franklin produced his newspaper, government-printing projects, and he took on a considerable volume of print for hire, such as forms, lottery tickets, handbills, and bookwork. Beginning in 1730, Franklin was printer of all paper money issued by Pennsylvania, New Jersey, and Delaware.

Perhaps one of his most recognized projects was *Poor Richard's Almanack*, which Franklin wrote under the pseudonym Richard Saunders and printed for twenty-five years beginning in 1732. As Franklin explained in his autobiography, "I endeavor'd to make it both entertaining and useful, and it accordingly came to be in such demand that I reap'd considerable profit from it, vending annually near ten thousand…. I consider'd it as a proper vehicle for conveying instruction among the common people who bought scarcely any other books."

Although Franklin was not the author of the many clever aphorisms contained in the *Almanack*, it can be argued that these sayings have been passed down and are part of our speech today because of his work. Here are a few of them:

> *Early to bed and early to rise, makes a man healthy, wealthy and wise* (1735).
> *The rotten apple spoils his companion* (1736).
> *No gains without pains* (1745).

During his printing career, Ben Franklin made important contributions to the technology and practice of printing. He made notable improvements to printing press design and helped to set up paper mills in the south. He established in 1778 the first printers' association in America (later named the Franklin Society), established the first public library, is credited with the idea for the first American magazine, and was appointed the first Postmaster by the Second Continental Congress in 1775. Benjamin Franklin created a franchise business model that launched two-dozen printing establishments up and down the Atlantic Coast.

In his *An Apology for Printers*, Franklin elaborated a philosophy for the role of the printer in modern society: "Printers are educated in the belief that when men differ in opinion, both sides ought equally to have the advantage of being heard by the publick, and that when truth and error have fair play, the former is always an overmatch for the latter."

If printers printed only what they believed, "the world would afterward have nothing to read but what happen'd to be the opinions of printers."

5

William Blake
1757–1827

Relief Etching
London, England
1788

William Blake is well-known as an English poet and painter. During his lifetime, he was not recognized for his genius, but today, Blake is viewed as a significant and early figure of the Romantic period of European culture. He was one of the most brilliant, and eccentric, representatives of the creative movement in the late eighteenth and early nineteenth centuries that emphasized the free expression of the feelings of the artist in his works.

William Blake was born on November 28, 1757, in the Soho district of London. He was the third of seven children, two of whom died in infancy. His father was James Blake, a London hosier, and his mother was Catherine Wright Armitage Blake. Since William exhibited a love of art at an early age, his parents enrolled him in drawing classes and elected to educate him on other subjects at home.

At age fourteen, William was apprenticed to the engraver James Basire of London for a seven-year term. At the end of this training, Blake emerged as a skilled engraver, but he chose to enroll as an art student at the Royal Academy instead of pursuing a professional career at this time.

In his years at the Academy, William Blake developed his creative style in opposition to the trends of the time, especially the popular extravagance of the Baroque painter, Peter Paul Rubens. Blake was drawn to the classical period and the style and precision of Michelangelo and Raphael. Some of Blake's early paintings were exhibited at the Royal Academy between 1780 and 1808.

William's gift of verse was also evident in his youth. A volume of his first poetry called *Poetical Sketches*, published by friends in 1783, contains lines that he had written as early as 1768 when William was only ten or eleven years old.

Throughout his life, beginning at age four, William claimed to have experienced visions. These apparitions were often of a spiritual and religious nature and formed the basis of his literary and visual creations. It was this mystical behavior that earned him a reputation of being an unstable man among his contemporaries. In fact, the noted poet William Wordsworth is quoted in a biography of Blake as having said, "There is something in the madness of this man which interests me more than the sanity of Lord Byron and Walter Scott."

William Blake was married to Catherine Boucher in 1782 in St. Mary's Church, Battersea. Although Catherine was illiterate—she signed their wedding contract with an "X"—William taught her to read and write and later trained her as an engraver. Catherine proved to be invaluable to him, helping to print his illuminated works and maintaining his spirits through several personal crises.

*****

Due to his posthumous acclaim in the literary and visual arts, Blake is not as well-known for his work as an engraver and his innovative contributions to the graphic arts. Aside from the spectacular beauty of his print work—his hand-illustrated series of epic and lyrical poems "Songs of Innocence" (1789) and "Songs of Experience" (1794) being among the most original and stunning prints ever produced—William invented the technique known as relief etching. This is the method that is associated with the illuminated printing of his most important work.

The previous method of engraving exposed the images and text to acid, and, therefore, the copper plate was recessed in those areas, and the transfer of ink to paper took place with the intaglio method. Beginning in 1788, at the age of thirty-one, Blake began his experiments with relief etching where the text of the poems was applied to copper plates with pens and brushes using an acid-resistant medium. He then etched the plates, dissolving the untreated copper and leaving the design to stand in relief. The pages printed from these plates were hand-painted in water-colors and then stitched together to finish the book.

Since he lacked popularity during his lifetime, Blake took to commercial work to make a living. Some of the books he illustrated are *Night Thoughts* by Edward Young, *The Grave* by Robert Blair, *Paradise Lost* by John Milton, *The Book of Job* and *The Pilgrim's Progress* by John Bunyon, and *The Divine Comedy* by Dante. The last two of these remained unfinished when Blake died in 1827.

Among his original works, some of the more popular titles, are *The Marriage of Heaven and Hell, Visions of the Daughters of Albion, The First Book of Urizen, Milton: A Poem*, and *Jerusalem: The Emanation of the Giant Albion*. The preface to the last of these contains the well-known verse "And did those feet in ancient time…" that was later composed as a hymn called "Jerusalem" and became a British national anthem.

William Blake died in his house on August 12, 1827, at age sixty-nine during his work. Living in near poverty, Catherine borrowed the money needed for the funeral of her loving husband. The service was attended by a small group of his closest friends. Today, a monument marks the approximate location of the remains of William Blake and his wife—Catherine died four years later—at Bunhill Fields in London.

William Blake lived at a time of great transformation in society, the beginnings of the industrial revolution. To some extent, his work reflects a deep emotional reaction to the coldness of those times and a longing for a simpler and more humanitarian world. However, despite his justified anger toward the socially negative impact of science and technology at the turn of the nineteenth century, Blake contributed to art in its literary, visual, and, some would say, even musical forms.

Had Blake lived in our day, he would no doubt have found in the technology of the personal computer an outlet for his visionary, mythical, and humanistic creative expression. As it happened, William Blake was one of the first and most important multimedia artists that ever lived.

6

Nicholas-Louis Robert
1761–1828

Papermaking Machine
Paris, France
1798

Industry experts say there are more than twenty thousand uses of paper in the world today. Communication, currency, hygiene, manufacturing, packaging, and construction are a few of the commercial applications of paper. Remarkably, even with the growth of electronic and digital alternatives, the worldwide production of paper and cardboard has continued to expand.

Paper is so ubiquitous in our lives that the subject of its origins and development would be of interest to almost anyone. Yet the versatility and utility of paper are such that we barely see it. As librarian and historian John Bidwell has said, "If you're noticing paper, that probably means there's something wrong with it."

Throughout its history, paper has served as the vehicle upon which an image is presented; it is the medium that delivers the message. The more invisible its "negative" form, the more effective paper carries its "positive" content.

Perhaps this background role is at least partially responsible for the perception that paper is more effect than cause, that the development of paper has always been driven by the requirements of what is printed upon it. However, the truth is that the evolution of paper and the methods of its making have proceeded in reciprocal relation to print as both impulse and consequence.

And so it was that paper was invented by Ts'ai Lun (Cai Lun) in China in AD 105 initially for the purpose of facilitating calligraphy. By the time Oriental woodblock printing was developed a century later, the making of handwriting papers was an established craft. The new printing method was adapted to the lightweight Chinese stock, and the soft impression of the woodblock prints could only be effectively made upon one side of this paper.

More than one thousand years later—after papermaking made its way across Asia and the Middle East and became established in Europe—a different kind of paper was being made. Crafted from macerated linen and cloth fibers, dipped in animal gelatin and dried to accommodate the quill pen, this stiff paper was used by Gutenberg in the application of his movable metal-printing types. The pressure of the inked metal-type impression took to these hardened European papers in a manner that facilitated printing on both sides of the sheet.

Four-and-a-half centuries later, as print was being transformed from a craft to an industry, paper once again emerged as the driver of innovation. Many years before the application of metal construction, cylinders, and steam power to printing machines, the technology of papermaking had undergone its own revolution.

In 1798, Frenchman Nicholas-Louis Robert began the industrialization of printing by inventing the papermaking machine. Although there were many technical hurdles to be overcome, the essential features of Robert's first successful invention have remained standard in paper manufacturing to this day.

Robert's technical innovation was the rotating cloth screen (wire) belt that received a continuous flow of fibers and delivered an unbroken sheet of wet paper to a pair of squeeze rollers. As the long strip of paper came off the machine, in Robert's configuration, it was hung by hand to dry on a series of bars or cables. This would later become a roll of paper.

Prior to Robert's invention, all paper was made by hand and consisted of dipping a framed mold with a porous surface into a vat of fibers suspended in water. Upon lifting the mold from the vat, a thin layer of fibers rested on top of the screen and was dried to form a sheet of paper. With this method, the mold could be used again only when the sheet had dried and been removed from it.

Aside from its important technical features, the key results of the papermaking machine were the following:

- Productivity: The volume of paper that previously would have taken many hand papermakers and many hours could be produced by a single operator in far less time.
- Paper Size: Sheet dimension was limited with handmade paper. It was not possible for a craftsman to adequately balance the fibers on top of a large mold. With Robert's machine, paper size was limited only by the width of the machine; the length was endless.

*****

Nicholas-Louis Robert was born in Paris in a small house on December 2, 1761. He had health problems as a child and was quite frail. Despite his condition, when he turned fifteen, the boy tried to join the French army because he was concerned that his aging parents could not afford to take care of him. The military did not allow him to join, and Nicholas-Louis was returned to his parents' home to continue his studies.

Four years later, following a period of severe mental anguish for having been a burden to the family, Nicholas-Louis again applied for military service. On April 23, 1780, he was admitted to the First Battalion of the Grenoble Artillery and was stationed in a garrison at Calais, a port city in Northern France across the English Channel from the cliff-top town of Dover.

Robert's military experience took a dramatic turn in 1781 when the young soldier was sent to war in the Caribbean during the American Revolution. The French Metz Artillery regiment sailed to Santo Domingo to fight an engagement against the British. Having won the battle, Nicholas-Louis returned home and shortly thereafter left the military at age twenty-eight to seek an occupation in civilian life.

With a mechanical aptitude, Robert turned to the printing craft and landed a clerkship with the well-known printing, type founding, and publishing company called Didot in Paris. After working for several years in the office, Nicholas-Louis sought a new opportunity in related industries and came upon the paper-mill operations of Francois Didot in Essones, a well-known papermaking center south of Paris.

The Didot papermaking operation was quite important as much of French paper currency was printed there. Nicholas-Louis initially obtained a position as mill personnel inspector. Making regular contact with the staff, Nicholas-Louis found that there were many antagonisms among the hand papermaking tradesmen, and these problems frustrated him.

After working in the mill for months, it occurred to Robert that there was a more advanced method for making paper, that it might be possible to get around the constant quarrelling among the paper-making trades. While the discord among the employees of the mill may have been the impulse for Robert's research—he was after all under the direction of the owner Didot—the fact is that his project led to a significant reduction in the cost and a more abundant supply of paper.

Robert's initial attempts at a papermaking machine were failures and termed "feeble" by Didot. Nonetheless, Didot encouraged Nicholas-Louis to continue with his work. For a time, the young man gave up on his project and became involved in other areas of mill operations. But under the constant prodding of his boss, he returned to his research in mechanical papermaking.

Soon, with the help of others at the Didot establishment, Robert saw that the basic principles of his original concept were sound. His earlier work was revived, and he proceeded with the construction of a device that was larger than the first machine. Finally, when trial sheets of paper had been effectively produced on the new machine, Didot encouraged Nicholas-Robert to file a patent for his invention.

On September 9, 1798, Robert and Didot traveled to Paris and presented a patent application to the French Minister of the Interior. In a letter that accompanied the application, Nicholas-Louis wrote, "It has been my dream to simplify the operation of making paper by forming it with infinite less expense and, above all, in making sheets of an extraordinary length without the help of any worker, using only mechanical means…. The machine makes for economy of time and expense and extraordinary paper, being 12 to 15 meters (472 to 590 inches) in length, if one wishes."

The patent was granted at a cost of 1,562 francs and dated January 18, 1799. Because the maximum length of handmade paper was about 32 inches, Robert's machine represented a significant step forward. The French government recognized the importance of his invention and dispatched an engineer to the Didot mill to assist in the construction of an improved model.

The Bureau of Arts and Trades wrote of it, "This machine forms paper of great width and indefinite length. The machine makes paper of perfect quality in thickness and gives advantages that cannot be derived from ordinary methods of forming paper by hand, where each sheet is limited in size in comparison with those made on this machine."

Mention must be made of the fact that Robert's papermaking machine was invented during the tumultuous years of the French Revolution (1789–1799). It is no accident that this technical break-through coincided with the spread of the ideas of "liberty, equality., and fraternity" that depended greatly upon the printed word.

However, as the historian Dard Hunter described, the "disturbed conditions" of the period were such that very little progress was made by Didot and Robert beyond the initial invention. As Robert became preoccupied with the financial benefits of his accomplishment, he sold his patent to Didot for 25,000 francs, a modest sum considering the implications of the invention.

Didot immediately sought to move further development of the project out of France and into the more stable environment of England. Didot's brother-in-law, John Gamble, was an English paper-mill owner, and the two corresponded to have a much larger machine constructed. It was then that the London stationers, Henry and Sealy Fourdrinier, became involved in building papermaking machines.

After several modifications to the wire, the Fourdrinier brothers invested sixty thousand pounds in the construction of a large machine that could make paper on continuous rolls. This attempt was eventually proven successful, and they were granted an English patent for it in 1806. However, the project was not commercially successful, and it bankrupted the Fourdriniers. Nevertheless, to this day—although its basic design was invented by Nicholas-Louis Robert—the papermaking machine bears the Fourdrinier name.

As the Fourdrinier machine underwent rapid development in the nineteenth century, the brilliant Nicholas-Louis was living quietly in France. When the first commercial papermaking machine was installed there in France in 1811, Robert was approaching fifty years of age and on his way out of the papermaking business. In 1812, he opened a small primary school in Vernouillet, northwest of Paris, where he worked as a poorly paid teacher. Never realizing the financial benefit from his invention, Nicholas-Louis Robert died broken and destitute on August 8, 1828. A monument to his memory was erected in 1912 outside the church at Vernouillet.

7

Alois Senefelder

1771–1834

Lithography
Bavaria, Germany
1798

By some estimates, offset lithography represents approximately two-thirds of all print today. Even with the rapid growth of digital printing, the oil-and-water-based process that also transfers the image to the paper with rubber blankets, it is still by far the most dominant form of print media production.

When separated from its offset component, lithography (which means "stone printing") has been in existence for more than two centuries. It is distinct from its relief, gravure, screen, xerographic, and ink-jet cousins in that the ink-carrying print image area is chemically separated from the nonimage area. It is this quality—both the positive and negative image are on the same flat surface—that places lithography in a category of printing technology all by itself called planography.

Although offset lithography is print's premiere technology (in terms of versatility and volume), it has only occupied this position since approximately 1950. The previous dominant technology—the relief letterpress method associated with Gutenberg—held that position for five hundred years.

We are fortunate that the inventor of lithography, Alois Senefelder, left behind a book with many details of his life, an explanation of how his discovery was made, and the methods for its effective use. The original translation of this volume, *The Invention of Lithography*, into English was made in 1911, and the Graphic Arts Technical Foundation (GATF) reprinted it on the bicentennial of Senefelder's accomplishment in 1998.

Alois Senefelder was born on November 6, 1771, in Prague where his actor father was appearing on stage at the time. The family lived in Munich, and this was where the young Alois attended school. He later won a scholarship to study law in the Bavarian city of Ingolstadt.

At the age of twenty, Senefelder's father died, and the young man left his studies at law school to support his mother and eight siblings. His interest in the theater and acting led him to writing plays as a way of earning money. Several plays that he had written as a teenager had received popular acclaim.

Initially, Alois went to a local printer with his manuscripts to get his work published. He quickly learned that his supplier had difficulty meeting deadlines, and he decided to move the work elsewhere. By the time his copies were available and bought by a bookseller, Senefelder discovered that his costs were barely covered.

Being ambitious, Senefelder decided, "I found that it would not be hard for me to learn and could not withstand the desire to own a small printing establishment myself." Lacking the resources to buy a printing press, the types, and paper, Senefelder engaged in various experiments with different etching and stereotype casting techniques.

When he was a student, Senefelder had studied chemistry. He was deliberate in his experimentation and tried many different methods of engraving using turpentine, wax, and tallow soap, as well as copper and zinc surfaces. He also worked with different ink formulas to get his image transferred effectively onto the paper.

Although it is a popular myth that Senefelder invented lithography entirely by chance, he explained, "I have told these things fully in order to prove to the reader that I did not invent stone printing by happy accident but that I arrived at it by a way pointed out by industrious thought."

He began working with stone initially for the purpose of "rubbing down my colors on it" and later to practice writing. Most of the methods he was experimenting with required that writing be done in reverse. It was at this time in 1796 that his discovery was made when he wrote a laundry list on a stone that he had prepared and found that the image could be inked and transferred to paper.

By 1798, the full process had been perfected and, on September 3, 1799, Senefelder was granted an exclusive license for it. He joined with the André family of music publishers and refined both the chemical processes and the special form of printing press required for using the lithographic stones. Senefelder called it "stone printing" or "chemical printing," but the French name "lithography" became more widely adopted.

Senefelder was recognized by King Maximilian Joseph of Bavaria and provided with a pension. A statue of Senefelder stands in the town of Solnhofen where lithographic stone is still quarried.

Alois Senefelder's contribution to printing was significant in that it was a process that was more affordable and could be more widely used. Although letterpress remained the dominant form of printing text, lithography became the preferred technique for art and graphic image reproduction. It would be much later, after the offset printing technique was added to the process, that lithography would come to dominate in the field of newspaper, book, and magazine publishing along with numerous types of commercial print.

8

Charles Stanhope
1753–1816

Iron Printing Press
Kent, England
1800

Historians generally agree that the first industrial revolution took place between 1760 and 1840. Among the features of this great economic and social transformation were (1) the progression from predominantly rural to urban society, (2) the replacement of handicraft with machine production, (3) the introduction of iron and steel in place of wood, and (4) the substitution of muscle power with new energy sources like coal-fired steam power.

A unique set of circumstances—a stable commercial environment, advances in iron making, and an abundance of skilled mechanics—made Britain the birthplace of the industrial revolution. Beginning with new techniques in textile production, industrial innovations spread rapidly to other manufacturing sectors and then across national borders in Europe and around the globe. All aspects of life would be touched by industrialization: population, politics, trade and commerce, science and culture, education, transportation, and communication.

It was during this era of remarkable change that the English aristocrat Charles Stanhope invented, sometime around 1800, the first printing press constructed wholly of iron. Prior to Stanhope's achievement, the design and build of printing machines had not changed in the three and a half centuries since Gutenberg.

Previously, small adjustments had been made to the wooden press. These related to structural stability, increased sheet size, and automation to reduce human muscle power. But even with the inclusion of some iron parts, the basic design of printing presses remained as they were in 1440.

With the Stanhope handpress, both the design of the impression mechanism as well as the material from which the machine was built were transformed; Stanhope's contribution was a crucial preliminary step in the industrial development of print communications.

Charles Stanhope, third Earl Stanhope, was born on August 3, 1753, the younger of two sons of Philip Stanhope, second Earl Stanhope, and his wife, Lady Grisel (Hamilton) Stanhope. As a member of the English peerage system—with titles like duke, earl, and baron—Charles is often referred to as Lord Stanhope or Earl Stanhope. Born into the English aristocracy, he was afforded a privileged upbringing and, at the age of nine, was enrolled by his parents at the prestigious Eton boarding school.

In 1763, following the death at age seventeen of his brother, Philip, from tuberculosis, Charles became family heir. His parents decided that Charles's "health should not be exposed to the English climate or the care of his mind to the capricious attention of the English schoolmaster," and the family relocated to Geneva, Switzerland. At age eleven, Charles was enrolled at the school in Geneva founded on the principles of John Calvin and there studied philosophy, science, and math.

As a teenager, Charles was known to be a devoted cricket player, an exceptional equestrian, and a well-mannered young man who was admired by his peers. At age seventeen, Charles won a prize in a Swedish competition for the best essay, written in French, on the construction of a pendulum.

While Charles was accomplished academically in math and science, he was also known to have talents in drawing and painting. As a nobleman, Charles had obligations as a militia commander, and he developed a passion for archery and musket shooting. At eighteen, he won a competition and was crowned the best shot and so-called "King of the Arquebusiers."

By the time Charles completed his education in Switzerland, his parents decided to move the family back to England. According to a published account, as the family and its entourage left Geneva in 1774, "the young gentleman was obliged to come out again and again to his old friends and companions who pressed round the coach to bid him farewell and expressed their sorrow for his departure and their wishes for his prosperity."

During their five-month journey home to England from Switzerland, the family made a stop in Paris. Charles was welcomed and "esteemed by most of the learned educated men of the capital" over the prize he had won for his paper on pendulum design. He was developing an international reputation as an innovator.

Upon his return to England, Charles used his skills in mechanics to win election to London's Royal Society, a world-renowned club founded by King Charles in the seventeenth century to promote the benefits and accomplishments of science. At the age of twenty, Charles embarked on a series of self-funded experiments and inventions, and his interest in such matters continued throughout his life.

The most important of these were

- a method for preventing counterfeiting of gold currency (1775);
- a system for fireproofing houses by starving a fire of air (1778);
- several mechanical "arithmetical machines" that could add, subtract, multiply, and divide. These inventions were early forerunners of computers (1777 and 1780);
- experiments in steamboat navigation and ship construction which included the invention of the split pin, later known as the cotter pin (1789);
- a popular single-lens microscope, known as the Stanhope, that was used in medical practice and for examination of transparent materials such as crystals and fluids (1806);

- a monochord or a single-string device used for tuning musical instruments; and
- improvements in canal locks and inland navigation (1806).

Charles Stanhope became so well-accomplished in international scientific circles that he was befriended by Benjamin Franklin. The two spent time together during Franklin's visits to England prior to the American Revolution. They shared a mutual interest in electricity, and in 1779, Charles Stanhope published a volume entitled *Principles of Electricity* that corroborated through experimental evidence Franklin's ideas about lightning rods.

By 1800, as has often happened in graphic arts history, the environment became ripe for a major step forward in printing methods. Charles Stanhope—who had the desire, know-how, and resources to make it happen—stepped forward with a significant breakthrough.

Due to his many scientific publishing activities and democratic political pursuits—some of which concerned freedom of the press—Charles was very familiar with printing technology. Among his concerns were the cost of production, the accuracy of the content, the beauty of the print quality, and the importance of books for the expansion of knowledge in society.

*****

All letterpress technologies require a means to transfer ink from the surface of the metal typeforms to the paper. This process requires the application of pressure—that is, an impression that mechanically drives the ink into the paper fibers. The pressure also creates a slight indentation in the shape of the letter forms in the surface of the paper.

Prior to 1800, press designs were based on the screw press that had been used for pressing grapes (wine) and olives (oil), cloth, and paper going back to Roman times. The screw mechanism is a complex arrangement of the screw, nut, spindle, and fixed bar that drives the platen—the flat plate that presses the paper against the typeform—downward. There are many historical drawings and engravings that illustrate how physical strength was required to pull the bar and make a printing impression with the Gutenberg-era press design.

Stanhope's innovation, according to historian James Moran, was that "he retained the conventional screw but separated it from the spindle and bar, inserting a system of compound levers between them. The effect of several levers acting one upon the other is to multiply considerably the power applied." The compound lever system was so successful that it became referred to as "Stanhope principles" and was incorporated into subsequent generations of hand-press design in the nineteenth century (Columbian, Albion, and Washington).

Other important Stanhope press changes were all iron construction including a massive frame formed in one piece, a double-size platen, and a regulator that controlled the intensity of the impression.

The Stanhope press would later undergo several important modifications, the most significant of which was strengthening the frame in 1806 to prevent the iron from cracking under the stress of repeated impressions. The second design, with its characteristic rounded cheeks, is what today is commonly associated with the Stanhope press.

The Times of London immediately adopted the Stanhope press, and it became successful across Europe and America in the first few decades of the 1800s. Meanwhile, further developments with all-iron handpresses would continue up to the end of the nineteenth century. However, driven by the rapid advancement of the industrial revolution, the next stage in the evolution of press design—the introduction of cylinders and steam power—would rapidly eclipse Stanhope's accomplishments.

*****

Charles Third Earl Stanhope was an unusual man. In addition to his many inventions and scientific studies, he devoted himself to radical political causes that often controverted his aristocratic background. He even sometimes referred to himself as "Citizen" Stanhope.

The origins of his democratic leanings were to be found in the influence of his father—who was a member of Parliament and an outspoken critic of the crown and proponent of habeas corpus—and his education in the radical environment of Geneva and the Revolutions in America (1776) and France (1789).

Known publicly as Viscount Mahon at the time, Charles was elected to Parliament in 1780 and adopted positions that conflicted with the political elite. His demands for electoral and finance reform and religious tolerance of dissenters and Catholics did not sit well with the establishment. Charles was also known to have campaigned against slavery and was party to the abolition bill known as the Slave Trade Act of 1807.

Charles Stanhope was an opponent of the war against the thirteen colonies and a supporter of John Wilkes, a British sympathizer of the American rebels. Despite his efforts on behalf of the oppressed and downtrodden in society, Charles Stanhope's personal eccentricities caused him, especially later in life, to be isolated from his family.

Always thinking of others before himself, he allowed his manse at Chevening, Kent, to fall into disrepair, and it is speculated that he had starved himself to death on a diet of soup and barley water. Charles Stanhope was interred "as a very poor man" in the family vault at Chevening Church one week after his death on December 15, 1816.

9

Friedrich Koenig

1774–1833

Cylinder Printing Press
London, England
1812

The period of the industrial development of printing is perhaps more resonant with our own time than any other era in the nearly six centuries since Gutenberg. At the turn of the nineteenth century, a fundamental transition began: press technology went from wood and hand power to iron and steam power; during the years 1800–1814, all the basic elements for the transformation of print from a craft to an industry were in place.

A measure of that revolution is the tenfold increase in press productivity in the first two decades of the nineteenth century:

Time Period	Up to 1800	1800	1818
Press technology	Gutenberg	Stanhope	Koenig
Sheets per hour	240	480	2400

While the Stanhope press is identified with the introduction of iron in press manufacture, the design is not a fundamental departure from the hand-operated, vertical-screw technique implemented by Gutenberg in 1440. Alongside iron, the key advancements in the industrialization of print were the cylinder and steam power. Patent and business records of the period identify the first press that had all these features with Friedrich Koenig.

Koenig was born on April 17, 1774, in the town of Eisleben, Saxony. As manufacturing society was emerging, the bookseller and printer shared with others the desire to mechanize the handpress. Koenig's first invention arrived 1803 with what was known as the Suhl press, a steam-powered system that had inking rollers. Due to the difficult economic conditions of early 1800s Germany—including lack of a functioning patent system—Koenig moved to London in 1806 to pursue his ambition as an inventor.

The Fleet Street printer Thomas Bensley, along with several other English investors, was convinced of Koenig's genius and agreed to finance his press experiments. Fellow countryman and engineer Andreas Bauer (1783–1860) joined Koenig in London, and they developed two new platen designs for which they achieved patents in 1810 and 1811.

The significant breakthrough came in 1812 with the invention of a steam-driven cylinder machine. Koenig wrote about it in *The Times*, "Impressions produced by means of cylinders, which had likewise been already attempted by others, without the desired effect, were again tried by me upon a new plan, namely, to place the sheet round the cylinder, thereby making it, as it were, part of the periphery."

Koenig's solution of the problem of effectively transferring a letterpress image to paper by means of a cylinder won him recognition. His method was first used in book production and later, after John Walter II of *The Times* joined the partnership with Bensley, in newspaper printing. At Walter's request, Koenig and Bauer built a double machine—being fed with sheets of paper from both ends—and obtained a patent for this device on June 23, 1813. The first issue of *The Times* printed with this technology was published on November 29, 1814.

John Walter II wrote in *The Times* on that day, "Our journal of this day presents to the public the practical result of the greatest improvement connected with printing since the discovery of the art itself... A system of machinery almost organic has been devised and arranged which, while it relieves all human frame of its most laborious efforts in printing, far exceeds all human powers in rapidity and dispatch… Of the person who made this discovery, we have little to add.… It must suffice to say farther that he is a Saxon by birth, that his name is Koenig, and that the invention has been executed under the direction of his friend and countryman, Bauer."

In 1816, Koenig and Bauer developed a perfecting cylinder press that was installed at Bensley's business. This system produced up to one thousand perfected sheets an hour. By this time, the two inventors wanted to sell their systems to the trade, but Bensley refused. Koenig and Bauer decided to return to Germany to manufacture presses under their own name, and on August 9, 1817, they founded their own firm in Oberzell near Würzburg, Germany.

Their company continued to play a significant role in the industrial evolution of printing machines as steam power gave way to electric motors and after the death of Friedrich Koenig on January 17, 1833. The company of Koenig and Bauer exists to this day (as KBA) at the same location and is the oldest printing press manufacturer in the world.

While European (both English and German) inventors dominated the early period of powered-cylinder presses, the center of gravity for printing technology shifted to America around midcentury, especially following the Civil War. The significant expansion of the US market and the emergence of conditions for experiment and manufacturing made it the ideal environment for further strides in printing-press development.

10

George Baxter
1804–1867

Pictorial Color Printing
London, England
1835

Prior to the invention of photography and photomechanical halftones, the printing of pictures required the handwork of skilled artists. For centuries, craftsmen used various manual techniques—engraving, etching, stippling, and drawing—to create original images on wood blocks, metal plates, or lithographic stones that could be inked and printed onto paper.

At each evolutionary stage—from relief to intaglio to lithography—pictorial printing became incrementally more productive. However, the craftsmen's work persisted, and the process remained slow. Well into the 1800s, it was common for creative work on pictures—such as engraved images printed alongside of text or images, printed separately from lithographic plates—to begin a year or two before the date of publication.

Although the artistic work was time-consuming, it often produced striking results. By the time color-printing methods were perfected, magnificent pictures began to appear. In the mid-nineteenth century, pictorial color printers—some using RBY or RBYK models and others using "tinting" techniques of up to twenty or even thirty different colors—were producing astonishingly beautiful prints.

As the methods advanced and cost per picture declined, the quantity of color printing grew. This expansion was in part due to the industrialization of printing-press machinery. The speed and volume of print was being driven up exponentially by metal cylinders, steam power, and rotary printing equipment. Another side of the surge in color was that more people than ever before had access to the new low-priced prints. For the first time, average people could buy printed copies of paintings and other original items previously seen only in the private collections of society elites. The color printer became for the people the disseminator of artistic masterpieces and the chronicler of contemporary events.

A few enterprising printers recognized that industrial society had created an opportunity to produce and sell pictures to the public. They established companies in cities and employed the available labor to print inexpensive color pictures for the growing urban population. In this environment, the Englishman George Baxter emerged as perhaps the most important figure of the era of pictorial color printing.

In some respects, the color work of George Baxter should be considered Victorian-age fine art. Even though he produced upward of twenty million prints during his lifetime, Baxter's work exhibits virtuosity in the technical aspects of platemaking and printing as well as his extraordinary gifts as an engraver. It would take many decades after his death for the meaning of Baxter's accomplishments to be fully appreciated. As C. T. Courtney Lewis explained in *George Baxter: Color Printer, His Life and Work* in 1908, "His genius was not unrecognized in his own day, yet it seems that it is only now that the hour of his complete triumph has sounded…he was not a printer merely: he was an artist, a pioneer, and a man of many and versatile talents."

The invention for which George Baxter is known—and for which he applied on October 23, 1835, and received a patent on April 23, 1836—is a complex one. To produce long press runs of beautiful and economically viability color pictures, Baxter combined two previously existing printing techniques: steel or copper plate intaglio printing and woodblock relief printing. Baxter's novelty was that the first impression was printed in black or gray ink with an intaglio outline or "key plate," and then subsequent multiple layers of color tinting were printed with relief woodblocks. His convergence of intaglio and relief printing produced pictures that were significantly superior to anything printed with either process independently of one another.

As Baxter himself explained in his patent application, "My invention consists in coloring such impressions of steel- and copper-plate engravings and lithographic and zincographic printing by means of block printing in place of coloring such impressions by hand as heretofore practiced and which is an expensive process and by such a process producing colored impressions of a high degree of perfection and far superior in appearance to those which are colored by hand and such prints as are obtained by means of block printing in various colors uncombined with copper and steel lithographic or zincographic impressions."

Prior to Baxter, key plates had been used in the more labor-intensive process of color tinting by hand. In the case of the woodblock method (also known as chromoxylography), it was used previously by others without the preliminary step of the key plate. It is also true that a combination of the two methods had been performed a century earlier by the Englishman Elisha Kirkall but with nothing approaching the level of Baxter's perfection or economy.

A decisive aspect of what became known as Baxter's process was the remarkably consistent quality of the entire color printing run. This was achieved with tight registration—using four pins or "pointers" in the press to hold the paper in position from one impression to the next—and oil-based inks. The advent of improved brightness of pigments and the permanent quality of the oil-based inks gave Baxter's prints a superior color fidelity. Among the first examples of Baxter's method was the frontispiece of Robert Mudie's 1835 book, *The Sea*. What may seem today as a subtle change, the picture of a boat at sea during sunset carries a degree of precision and detail that was not achievable prior to Baxter's innovation.

George Baxter was born on July 31, 1804, at Lewes in the southeastern English county of Sussex. This area is known to have been a center of the early English printing and papermaking industries. George was the second son of John Baxter, the proprietor of a typography, printing, and publishing establishment in Lewes. George's father would gain in his lifetime a reputation as an advanced printer who was the first to test and perfect several early industrial innovations in printing-press technology.

George attended Cliffe House Academy and went to high school at St. Ann's in Lewes. After he finished school, he worked in a bookshop in Brighton, a seaside town less than ten miles from home. Later, although the record is unclear, he apprenticed as a wood engraver and lithographer. By the age of nineteen, George was focusing on the artistic elements of printing rather than the mechanical, and he began making a name for himself as a gifted illustrator. By 1826, it is known that George Baxter was in Lewes at his father's establishment identifying himself as a "wood engraver."

At age twenty-three, George migrated to London and set up his own business as an engraver and printer. Six months after starting his enterprise in London, George married Mary Harrild, daughter of Robert Harrild, a printing-industry innovator and business partner of John Baxter. The record shows that at this time, George Baxter began his experiments with color printing.

Once he had demonstrated to himself—if not also to everyone in the printing business—that his patented process represented an important breakthrough, Baxter started on a path that would continue for the next thirty years. During the period of his first patent grant (1834–1849), Baxter had no competitors in England for the process that he advertised as "Pictorial Colour Printing for Book Illustration and Picture Printing." At the end of 1836, Baxter produced a volume called *The Pictorial Album*, or *Cabinet of Paintings*, that was published by Chapman and Hall. This project, which contained ten pictures including reproductions of several works of well-known artists of the day along with a frontispiece, was the first major publication of Baxter's process. Some have said it was the high watermark of his craft.

In 1841, Baxter printed two pictures called "Her Most Gracious Majesty Receiving the Sacrament at Her Coronation" and "The

Arrival of Her Most Gracious Majesty Queen Victoria at the House of Lords to Open Her First Parliament" that include two hundred portraits of identifiable guests at the event at Westminster Abbey in 1938. The pictures, which were 21 3/4" × 17 1/2", were prepared in cooperation with Buckingham Palace and gained Baxter direct access to the queen and other royals in Britain and elsewhere in Europe.

Confident of the superiority of his methods, George Baxter was not shy about self-promotion as he gained a level of notoriety for his invention. However, as Baxter was continually preoccupied with the artistic and technical aspects of his business, he was never successful financially. In 1849, he petitioned the Privy Council and was granted a five-year extension on his patent on the grounds that he had lost money during the previous fourteen years.

In 1851, Baxter prints were on display at the Great Exhibition (also called the Crystal Palace Exhibition) in London. Of his work, the official catalog of the expo said, "Mr. George Baxter, the patentee of the process of printing in oil colours, exhibits in the Fine Art Court, upwards of sixty specimens (from the largest size to the smallest miniature), of his choicest productions… The visitors will indeed be delighted with these charming specimens which form the principal attraction in the Fine Art Court."

Baxter also exhibited prints at the international expos in New York City in 1852 and Paris in 1855. He was awarded medals for these entries. Later, Baxter produced a series of prints called "Gems of the Great Exhibition" which show the grandeur of his vision and the dexterity of his engraving skills.

In later years, while still holding his patent, Baxter took to licensing his method to other printers as a means of generating income. Having obtained color-printing patents in France, Belgium, and Germany, as well as Britain, Baxter sold annual licenses to a handful of printers in all these countries. Some have said that the work of these printers never approached Baxter's in graceful detail and delicate coloring. Six years after his process went into the public domain, Baxter liquidated his oil-color printing business and sold off his inventory of prints and intaglio plates and woodblocks to another printer. Part of this arrangement included Baxter's agreement to provide technical assistance to the new owner.

The reasons for his decision to exit the business are unknown. By the 1860s, other competing methods, such as chromolithography (Engelmann, 1837) and photography (Daguerre, 1839), were challenging the Baxter method in both quality and cost. By 1865, the remainder of Baxter's printing business went bankrupt. In late 1866, George Baxter was struck in the head during an accident involving a horse-drawn omnibus. He died at his residence in Sydenham on January 11, 1867, and was buried at Christ Church, Forest Hill in London. A red granite obelisk above his grave bears the inscription "the sole inventor and patentee of oil-color printing."

Some believed that Baxter's significance and contribution had been exaggerated by a cult of enthusiasm built up by clubs and associations organized to collect copies of his works. One such critic, R. M. Burch, wrote of Baxter in 1910, "Had he not been, rediscovered… his name and fame would in all probability have completely passed into the limbo of forgetfulness." However, George Baxter is remembered for the lasting impact of his original color-printing method and for making color printing popular and viable. Like others before and after him, Baxter's genius and creative gifts intersected with important changes in the means and methods of printing during his lifetime. It is undeniable that George Baxter played a decisive role in expanding the influence of print upon society during the Victorian era.

11

———

Richard March Hoe
1812–1886

Rotary Printing Press
New York, New York
1847

The transformation of printing from handcraft to industrial manufacturing began in 1812. Koenig and Bauer—through the application of iron, steam power, and, above all, the cylinder—replaced the wood construction, hand power, and screw mechanism of the Gutenberg press that had existed for the previous 370 years.

While these changes dramatically increased productivity, the industrialization of print machinery remained incomplete. Mechanical limitations, the consistency of papermaking manufacture and print quality issues prevented the cylinder press from reaching speeds greater than a few thousand sheets per hour. By midcentury, the demands of publishers for faster and larger print production volumes required another technological leap. In 1847, Richard March Hoe provided the solution with his invention of the first rotary printing press.

When reviewing the evolution of the press, an understanding of print fundamentals is important. Except for modern-day digital systems—where an infinitely variable image is rendered directly to the substrate by ink-jet print heads—all previous printing machinery had an important common feature: an image carrier. The image carrier is the surface from which ink is transferred to paper or other printing substrate.

In letterpress printing, the image carrier is the raised metal typeform. The letterpress process also requires an impression surface where pressure is applied to transfer the ink to paper. During the industrialization of printing, both the image carrier and the impression mechanism were converted from a flat platen to a curved rotary surface.

The Koenig cylinder press was a transitional or hybrid development between the platen press and the rotary press. With the cylinder press, the image carrier remained flat while the impression surface was curved. The ink was applied to the paper as it rode on the bed underneath the impression cylinder.

It took another thirty-five years of experience and experimentation before the typeforms could be mounted on a curved surface and the paper could move on its own between the impression and the "revolving-type" cylinders. This is the core achievement of the rotary press identified with Richard March Hoe.

There were many individuals and companies that worked on creating a rotary printing machine that did away with the platen typeform bed. As far back as 1790, an Englishman name William Nicholson took out a patent for a system that designed letterforms around a cylinder. However, the idea was never delivered practically.

The story of the first commercially successful rotary press is the story of the Hoe family. Richard March Hoe's father, Robert Hoe (1784–1833), was born in Leicestershire, England. With skills as a carpenter, Robert emigrated to the United States at age eighteen. He later formed with his brothers-in-law, Peter and Matthew Smith, a wooden press-manufacturing firm in New York City. In 1823, he gained sole ownership, and the R. Hoe & Company was founded, and it became the first company to build an iron and steam-driven cylinder press in America based upon the designs of Koenig and Bauer.

After Robert Hoe's death in 1833, his sons, Richard March Hoe and Robert Hoe II, took over the business. More than any other firm, the R. Hoe & Company expressed the fact that during the two decades between 1830 and 1850, the center of printing technology development had shifted from Europe to America.

With Richard March Hoe at the helm—and driven by the demand for the mass circulation and daily newspaper—the company developed an innovative technique for mechanically feeding sheets of paper into a press. Later, the company designed and built the "Hoe Type-Revolving Machine," its most significant invention and the first functioning rotary press. This system was so successful and productive that it became known as the "Lightning Press."

Working on this press in 1845–46—Richard March Hoe obtained patents for it in 1847—the problem of mounting type-forms onto the curved surface of a cylinder was solved. According to an account published in 1902 by Robert Hoe III, the grandson of the founder, "The basis of these inventions consisted in an apparatus for securely fastening the forms of type on a central cylinder placed in a horizontal position. This was accomplished by the construction of cast-iron beds, one for each page of the newspaper. The column rules were made 'V' shaped—that is, tapering toward the feet of the type. It was found that, with proper arrangement for locking up or securing the type upon these beds, it could be held firmly in position, the surface form a true circle, and the cylinder revolved at any speed without danger of the type falling out."

Robert Hoe III also noted the importance of these developments, "As the demands of the newspapers increased, more impression cylinders were added until these machines were made with as many as ten grouped around the central cylinder, giving an aggregate speed of about twenty thousand papers per hour printed upon one side. A revolution in newspaper printing took place."

With its six-and-a-half-foot diameter central revolving-type cylinder, these new machines were rapidly adopted across the US and in Britain. In 1858, *The Times* of London ordered two ten-cylinder presses to replace the Applegath vertical press that had previously been in use. R. Hoe & Company also built two-, four-, six- and eight-cylinder systems to accommodate the different needs of newspaper publishers.

The challenges that remained to be solved were developing a rotary press capable of printing on both sides (perfecting) and from a roll (web) of paper. These issues would be conquered after the Civil War, and R. Hoe & Company would play a role, along with other important figures, in their development.

12

William Bullock

1813–1867

Web Press
Pittsburgh, Pennsylvania
1863

As mentioned previously, a significant driving force behind the industrial transformation of printing was the expansion of newspaper publishing. In 1840 there were 1,300 newspapers in America; by 1850, this number had doubled, and most big cities had multiple daily papers. New York had fifteen dailies, Boston twelve, and Philadelphia and New Orleans each had ten.

By 1860, the telegraph and transatlantic cable increased the speed of news delivery, and the largest circulation paper in the world, *New York Herald*, was distributing seventy-seven thousand copies daily. When the Civil War began on April 12, 1861, the *New York Herald*'s circulation shot up to 107,000 and did not fall below one hundred thousand until after the war. By this time, there were about 3,700 newspapers in the US, and 387 of them were daily papers.

This was the beginning of the era of enormous publishing empires, and major investments were being made in press-technology development. The limitations of previous generation-powered rotary presses had to be overcome to satisfy America's population growth and the exploding demand for news in printed form. In 1835, the English inventor and pioneer of the modern postal system,

Sir Rowland Hill, suggested printing on both sides of a roll of paper. However, as has been the case with many previous printing press ideas, it is one thing to suggest an idea and quite another to execute it practically. There were several important technical issues involved in the first successful rotary web-printing press. The most important of these was the rapid cutting off and delivery of the paper either before or after it was printed.

The solution to the problem of web cutoff is widely recognized as being made by the inventor William Bullock in Pittsburgh, Pennsylvania, beginning around 1861. In the patent granted to Bullock on April 14, 1863, the inventor describes the significance of his achievement, "My improved machine for printing from moving type or stereotype plates belongs to that class of power-printing presses in which the paper is furnished to the machine in a continuous web or roll and by which the sheets are severed from the web, printed on both sides, and delivered from the machines thus 'perfected.'… In my machine, however, there is but one delivery apparatus, which is simple in construction and works as rapidly as the machine can be driven and the sheets printed so that by my invention the great obstacle to the rapid operation of the printing press is successfully overcome."

The prototype of Bullock's breakthrough system was installed at the *Cincinnati Times* in 1863, and the *Philadelphia Inquirer* acquired the first fully functioning model in 1865. The essential components of Bullock's rotary web-cutoff technology remain in use to this day in web presses all over the world.

William Bullock's biography is one of a harsh life and, ultimately, tragic death. Born in 1813 in Greenville, New York, and orphaned shortly thereafter, Bullock was apprenticed at the age of eight to a foundryman and machinist by his brother. By the age of twenty-one, William had his own shop and was working diligently as an inventor. He moved to Savannah, Georgia, in the late 1830s after inventing a shingle-cutting machine and where he also built hay and cotton presses. He returned to New York and made artificial legs and invented a grain drill, seed planter, and a lath-cutting machine. In 1849, he won second prize from the Franklin Institute in Philadelphia for his grain drill.

Bullock's interest in printing presses began in the 1850s. He was editor of *The American Eagle* in Catskill, New York, in 1853 and then moved to New York City to work as a mechanical engineer. He ended up building a high-speed press for the nationally circulated *Leslie's Weekly* in 1860. William Bullock moved to Pittsburgh in 1861, and it seems he aspired to become a patent attorney having considerable experience with patent filings. It was around this time that he changed his listing in a Pittsburgh business directory to "manufacturer of printing presses."

The original Bullock press design cut the paper off before printing on it. Later models of his machine cut the web after it had been printed upon, just as it is done today in most web offset presses. Later, a folder invented by Walter Scott and multiple webs were added to the system, further improving press productivity.

William Bullock did not live to see these additions to his invention. On April 12, 1867, he died from injuries sustained when his leg became entangled in the drive mechanism of a press he was installing at the *Philadelphia Public Ledger*. Bullock is buried in Union Dale Cemetery on Pittsburgh's North Side. He was belatedly recognized in 1964 as one of the city's geniuses with a bronze memorial marker that reads: "His invention of the rotary web press (1863) made the modern newspaper possible."

The industrial development of printing—much like our present digital, Internet, and social media era—transformed the landscape of information and news publishing and distribution. Those who studied and understood the form and content of the successive waves of tecshnology revolution were able to take advantage of the opportunities that presented themselves. Individuals like Friedrich Koenig, Richard March Hoe, and William Bullock exemplified the best of their respective generations as printing passed through a transformative period during the nineteenth century and the global center of technical innovation moved from Europe to America.

13

Christopher Latham Sholes

1819–1890

Typewriter
Milwaukee, Wisconsin
1867

At the corner of Fourth and State Streets in Milwaukee, a historical marker reads, "At 318 State Street, 300 feet northeast of here, C. Latham Sholes perfected the first practical typewriter in 1869. Here he worked with Carlos Glidden, Samuel W. Soulé, and Matthias Schwalbach in the machine shop of C. F. Kleinsteuber."

With Kleinsteuber's workshop long gone, the marker stands on the property of an arena which is the former home of the Milwaukee Bucks and several other professional sports teams. The 12,700-seat facility was built in 1968. The Bucks and the other teams moved across State Street to the newer 18,000-seat BMO Harris Bradley Center built in 1988. The modern surroundings of the typewriter's birthplace are a reminder of how much time has passed since Sholes's invention "freed the world from pen slavery." Fortunately, a surviving photo of Kleinsteuber's machine shop provides a glimpse into what life was like for Milwaukeeans between the Civil War and the automobile.

Christopher Latham Sholes was born February 14, 1819, in Mooresburg in Montour County, Pennsylvania, not far from the county seat of Danville. Sholes was born in a cellarless log house, eighteen feet square, a story and a half and with four windows.

After his family moved to Danville, Christopher's mother, Catherine (Cook) Sholes, died in 1826 when he was seven. His father, Orrin Sholes, was a cabinetmaker, and he had a workshop in town. While attending Henderson's school in Danville, Christopher worked in his father's cabinet shop. After graduation at age fourteen, he was apprenticed to the printing trade as a shop "devil" on the *Danville Democratic Intelligencer.*

By the time Sholes achieved master printer status at age eighteen, his family decided to move to Green Bay, Wisconsin. Encouraged to make the 750-mile trek by President Andrew Jackson's proclamation of public land sales, the Sholses were among the frontline of settlers who relocated to the Territory of Wisconsin. Christopher's older brother, Charles, had established himself as a printer and political figure in the area. Prior to the arrival of the family, Charles had become the publisher and editor of the *Green Bay Democrat.* The elder Sholes would go on, following Wisconsin statehood in 1848, to serve in both houses of the state legislature as well as mayor of Kenosha.

With a combination of his brother's influence and his own exceptional talents, Christopher was appointed official printer and took charge of the House Journal of the Wisconsin Territorial Legislature. At age twenty, he became editor of the *Wisconsin Enquirer,* a Madison publication owned by his brother. In 1840, Christopher moved to Southport (later Kenosha). He launched and became editor of the *Southport Telegraph.* The paper took its name from the invention of Samuel Morse. Sholes recognized the telegraph as a breakthrough communications technology that would improve the speed of news distribution. In 1844, he also became town postmaster.

Inevitably, like his brother, Christopher entered politics. He served two terms (1848–49 and 1856–57) in the state senate and one term in the assembly (1852–53). In 1860, Sholes moved to Milwaukee where he became postmaster and commissioner of public works. He was also at different times editor of the *Milwaukee Daily Sentinel* and the *Milwaukee News.* Sholes's inventive genius was sparked by business needs. His first invention was for printing the address of subscribers into the margin of newspapers, an early form of what we now call "variable data printing." He also worked with fellow inventors Samuel W. Soulé (machinist) and Carlos S. Glidden (attorney) on a machine for automatically numbering the pages of blank books and for sequentially numbering checks. Sholes obtained US patents in 1864 for these inventions along with one for a combination shoe brush, shoe scraper he invented with C. F. J. Moller in 1866.

By the 1860s, many people were interested in developing, investing in, or inventing a "machine for writing with type or printing on paper or other substance" as one such system was called. The race was on to see who could come up with a viable, personal, and portable alternative to the four-hundred-year-old relief-printing process associated with Johannes Gutenberg.

Attempts had been made to conceptualize and even produce a typewriter going back to the 1700s. Englishman Henry Mill received a patent from Queen Anne in 1714 that called for "impressing or transcribing letters singly or progressively one after another, as in writing, where all writing whatsoever may be engrossed in paper or parchment so neat and exact as not to be distinguished from print..." However, as promising as it sounded, Mill left behind no drawings or record of any existing machine to go along with what was a breakthrough idea.

Sholes was inspired to solve the technical riddle of the typewriter after he saw the July 6, 1867, issue of *Scientific American.* The SA article reported an invention that had been exhibited at the London Society of Arts by John Pratt (later known as the Pterotype or "winged type") of Centre, Alabama.

The SA editors captured the implications of what would become later Sholes's invention: "Legal copying and writing and delivery of sermons and lectures, not to speak of letters and editorials, will undergo a revolution as remarkable as that effected in books by the invention of printing, and the weary process of learning penmanship in schools will be reduced to the acquirement of the art of writing one's own signature and playing on the literary piano above described or rather on its improved successors."

Sholes vision for the typewriter was a natural extension of his numbering machine inventions of 1864. The proof of concept was a primitive system of wood frame, glass platen, and brass bar attached to a Morse telegraph "key." It produced the letter "w" repeatedly by striking through a piece of carbon onto a sheet of paper against the glass. After receiving an enthusiastic response from those who saw the concept, Sholes worked with Soulé, Glidden, and the engineer, Matthias Schwalbach, throughout the summer and fall of 1867 to develop the first working typewriter. This machine used a keyboard that looked like that of a piano, and it had a typewriter ribbon to transfer the image to paper. On June 23, 1868, Sholes, Glidden, and Soulé received a patent for the design, and it is recognized as the first practical typewriter.

The innovations represented by this invention are too numerous to explain in detail here. As the patent, filed on October 11, 1867, explains, "Its features are a better way of working the typebars, of holding the paper on the carriage, of moving and regulating the movement of the carriage, of holding, applying, and moving the inking ribbon, a self-adjusting platen, and a rest or cushion for the typebars to follow." However, the inventors acknowledged the advances of others before them and filed the patent for "improvement in typewriting machines," not the invention of the typewriter itself.

Mention should be made here of the QWERTY keyboard about which much has been written. The record shows that the keyboard design underwent an evolution beginning with a straightforward listing of numbers and letters of the alphabet. Upon testing and subsequent design improvements—coinciding with the collaboration of the inventors with investor James Densmore—it was seen that frequent jamming of the typebars was a barrier to practical use of the machine.

Sholes worked with Densmore's brother, Amos, an educator, to make a statistical analysis of the most frequently used letter combinations. From there, Sholes changed the keyboard design such that common letter pairs were separated by a "lag time," and the instance of typebar jams was reduced. The resultant QWERTY keyboard remains in use today even though these mechanical considerations are no longer present.

Something must also be said of Sholes's character. While he was a man of significant talents and influence—Sholes left his position as editor of the *Milwaukee Sentinel* to accept an appointment by President Abraham Lincoln as collector of the Port of Milwaukee—he was also a man of great principles and humility. An active opponent of slavery, Sholes was an abolitionist and founding member of Lincoln's Republican Party. He supported Joshua Glover's case challenging the Fugitive Slave Act in 1854.

Historians universally recognize the magnanimity of Sholes and his preoccupation with progress over personal gain and recognition. It is a fact that he sold his invention to Remington (the Civil War gun manufacturer) for mass production and gladly accepted a one-time payment of $12,000 instead of a royalty contract.

Christopher Latham Sholes suffered throughout his life from persistent health issues that were likely the product of the harsh conditions of his upbringing. He died on February 17, 1890, after a long bout with tuberculosis and was buried in an unmarked grave in Milwaukee's Forest Home Cemetery. An effort was mounted in the early twentieth century to appropriately recognize Sholes, and his grave was marked with a monument. It says, "Dedicated by the young men and women of America in grateful memory of one who materially aided in the world's progress."

14

Frederic Ives
1856–1937

Halftone Photo Reproduction Process
Ithaca, New York
1881

Frederic Eugene Ives's inventions and discoveries in the field of visual communications technology—the development of the first halftone reproduction process being the most significant—span six decades and are among the greatest contributions of any individual to the industry. The son of a farmer-turned-country storekeeper from a small town in rural Connecticut, Ives developed an interest in printing when he found a small handpress in his father's shop. He left school before the age of twelve to find a job and earn a living after the early death of his father from pulmonary consumption. More than a year later, young Frederic obtained an apprenticeship in the printing offices of the *Litchfield Enquirer* where he earned the statewide reputation among newspaper printers as "the natural printer." This was by virtue of the superb quality of his work.

Lacking any formal education, Ives developed a passion for investigation and experimentation in photography and engraving while working late into the night in the attic of a building directly across the village green from the *Enquirer* office. After completing his apprenticeship at the age of seventeen, he became a journeyman job printer for a printing establishment in Ithaca, New York, some two hundred miles away from his hometown. This was followed a year later with an application for a job running the photographic laboratory at Cornell University. After initially being declared too young and inexperienced for the position, Ives was selected by the university administration for employment on a "trial basis."

While Ives's tenure at Cornell lasted just four years, it was during this time that he would go on to develop some of his most important ideas; ideas that would transform the world of printing. His invention of the halftone photoengraving process in 1881 and later the crossline screen for direct photographic halftone reproduction stands out as a transition period in the history of printing and publishing. Ives had created for the first time the technology and method for reproducing with ink-on-paper printing processes all the tonal values and richness of detail from an original photographic image. Prior to this discovery, imagery in print was confined to the highly skilled and time-consuming efforts of handicraft wood engravers and resembled works of art more than living scenes as perceived by the human eye.

In its essential features, the halftone process remains in use today as the most common method for photographic reproduction in print. It is safe to say that the offset lithographic process, the predominant printing technology of the past half century, could not exist without Ives's invention. Each day, millions upon millions of printed products—newspapers, books, magazines, brochures, calendars, wrapping paper, greeting cards, packaging materials, billboards, to name only a few—are produced by machinery that utilizes what was once known as the "Ives process." Simply put, the halftone is an optical illusion: small dots of various sizes that are equidistant from each other create the appearance, at an appropriate viewing distance, of continuous gradations of tone. Since many printing processes can only transfer a solid film of ink to a sheet of paper (or other substrate), the halftone is the most effective method for reliably simulating a continuous tone image such as a photograph. Measured in lines per inch, the halftone screen is the essential building block of the printed page upon which everything else depends.

Ives also made major contributions to the development of color photography and microscopy. Among his seventy patents were the photochromoscope camera, the chromogram, and the single-objective binocular microscope. In his later years, when asked how he came to devote himself to the field of optics without what was considered the requisite mathematics and physics training, Ives quoted Robert Louis Stevenson's remark about his father, the lighthouse engineer, who he said had a "sentiment for optics."

An unusually gifted man, Ives wrote about himself in his *The Autobiography of an Amateur Inventor*, "The writer belongs to a period when some of the most revolutionary inventions were made by men not specially trained for such work but were impelled to undertake it by the possession of what Sir William Abney once termed 'instinctive genius.' To this class of men, I would apply the term 'amateur inventors.'… Some men are as naturally inventors as others are poets, fiction writers, statesmen, or merchants, and the typical amateur inventor will pursue his course through any amount of poverty and hardship and indifference, thinking much more about his work than about any material reward which it might bring."

Grist! Are your messages to the world as *speedily* printed as they should be? A message from you to us may bring some surprising thoughts on this important subject. Send it now. ❦❦❦ From out of the whirling wheels of this small utterer of well-printed sheets comes a great volume of the world's grist of letters, forms, plans, designs, etc. All are microscopically accurate duplicates of their originals. But the fact that the Mimeograph will deliver thousands of duplicate letters within the hour of dictation establishes its supremacy in the world of action. It is an hour-saver—as it is a dollar-saver. Your message—for booklet "M"—now! A. B. Dick Company, Chicago—and New York.
MIMEOGRAPH
EDISON-DICK

15

Albert Blake Dick

1856–1934

Mimeograph
Chicago, Illinois
1884

As late as the early 1970s, schoolteachers gave homework and classroom assignments, quizzes, and tests on Ditto worksheets. They were used so often that students became intimately familiar with the aniline purple color of the Ditto, as well as the mesmerizing smell that emanated from the freshly printed sheets.

Making Dittos was a two-step process. The first step was to prepare the master, a two-ply form that had an easy-to-write-on paper sheet on top and a wax-coated sheet on the bottom. Teachers would either handwrite or typewrite the schoolwork onto one of these typically letter-size Ditto master forms. The pressure of the pen or the typewriter would transfer wax from the bottom sheet onto the back of the top sheet.

The second step—after discarding what was left of the bottom sheet—was to mount the master, bottom side up onto the Ditto duplicating drum. The wrong-reading wax image contained the "ink" that was progressively broken down by the chemical spread across the drum as it was rotated—often by cranking the cylinder manually—and met the paper. Several dozen Ditto sheets could be easily produced within minutes.

The Ditto machine was the American variety of a duplicating system that became popular internationally—the Banda in the UK and the Roneo in France and Australia—in schools, churches, clubs, and other small organizations. The Ditto is known generically as a spirit duplicator; the term "spirit" refers to its alcohol-based solvent.

The faintly pleasant odor of the Ditto came from the fact that each sheet was essentially being coated with "10 percent of mono-fluoro trichloro methane and 90 percent of a mixture of 50 percent methyl alcohol, 40 percent ethyl alcohol, 5 percent water, and 5 percent of ethylene glycol mono-ethyl ether." This composition was developed in the 1930s as a less dangerous alternative to the original spirits of pure methyl/ethyl alcohol with a tendency to combust in confined spaces and air temperatures above 100°F.

Since spirit duplicators were limited to a maximum of about three hundred copies per master and the quality of reproduction as well as the cost per copy were very low, they became a DIY alternative to more sophisticated printing equipment. The Ditto was perhaps the most successful small office-copying system during the four decades prior to the ascension of xerographic toner-based photocopiers in the 1970s.

Spirit duplicators were one of several document reproduction technologies that were developed for the office rather than the printing plant. Office duplicators were first invented in the late 1800s in response to the demands of business for efficiency and economy in reproducing company documents in small numbers. Alongside the typewriter, office duplicators answered the problem of business forms and letters by replacing the tedium of copying each one by hand.

With commercially available printing machinery was very costly and too slow for these on-demand and short-run copying needs, an alternative had to be found. In 1884, a Chicago lumber business-man devised a stencil-based method of document duplication that he would later call the "mimeograph." From that point forward, the name A. B. Dick has been associated with the duplicating era of print technology.

Albert Blake Dick was born on April 16, 1856, in Galesburg, Illinois, a town about 175 miles southwest of Chicago and 50 miles northwest of Peoria. His parents, Adam Dick and Rebecca Wible, were from western Pennsylvania and decided to settle in Galesburg after helping to establish a church congregation in Quincy, Illinois. Albert attended public school in Galesburg and then went to work for a farm-equipment manufacturer in the area. After showing success as a manager, he became a partner in a lumber company. Just shy of his twenty-eighth birthday on April 11, 1884, the young Albert incorporated a lumber firm, the A. B. Dick Company, located at 740 Jackson Boulevard in Chicago.

It was during these early days that Albert preoccupied himself with the problem of business document reproduction. He rebelled against the effort wasted daily by hand copying price lists. Albert spent many hours experimenting with many unsuccessful ideas, most of them using the stencil principle. The stencil method is distinct from other printing methods in which an inked image is mechanically transferred onto a substrate. Once a stencil sheet is prepared, it is mounted upon the ink-filled rotary duplicating drum. When a blank sheet of paper is brought into pressured contact with the rotating drum, ink is forced through the holes in the stencil onto the paper. Silk screening is also a form of stencil printing, but it utilizes a flatbed and squeegee process that is more wasteful than the process associated with A.B. Dick.

"My aim," Albert would describe on the fiftieth anniversary of his company, "was to find a new means of duplicating letters other than by printing from movable types, something more economical of both time and money." It did not take long. Sometime within the first year of his lumber firm, Albert sat down at his desk, and across a piece of waxed paper, he forced an awl (long-pointed metal spike). After looking more carefully at what he had done, Albert noticed that the awl had left a series of tiny perforations on the wax paper. Developing this method, he perfected a sufficiently coated wax sheet as well as a stylus with which to write that could enable enough ink to be transferred to blank sheets of paper.

While his invention had achieved the immediate goal that he had set for himself, Albert returned his attention back to the development of his lumber company. For the next three years, the stencil-du-

plicating technique he pioneered remained an entirely internal matter at the A. B. Dick Company. In 1887, following multiple inquiries
by outsiders as to where a device such as his could be obtained, Albert
decided to patent his invention with a plan to market and sell it to
the broader business community. In a most peculiar and fortuitous
coincidence, it turned out that Thomas Alva Edison already held the
patent for Albert's stencil-duplicating concept. In 1876, Edison had
obtained a patent for his "Edison Electric Pen," a more primitive
implementation of the same principles that Albert had discovered
independently but nonetheless a very popular product

Rather than walk away from the opportunity, the young Albert
Blake Dick decided to approach Edison with his superior idea and
see what arrangements could be made. Edison, ever the entrepreneur, readily accepted that Albert's solution was simpler and more
economical than his motorized pen technology. Furthermore, Edison
agreed that his name would be associated with the product that
Albert would develop and market. In preparing to manufacture and
sell the stencil system, Albert developed the trademark name. As he
explained in 1934, "One day, an old friend hit upon the combination of 'mime' and 'graph.' But it didn't have the right swing. It wasn't
euphonious. Then the 'o' was added to give it the swing, and the
right euphony was acquired."

The original Model 0 Flatbed Duplicator was sold as the Edison
Mimeograph in 1887 and cost $12. A.B. Dick's inventive genius
did not stop there. By 1900, the company had developed the rotary
Edison Diaphragm Mimeograph No. 61, the Edison Oscillating
Mimeograph No. 71, and the A.B. Dick No.1 Folder, an automatic
letter-folding machine. By the 1910, there were two hundred thousand mimeograph machines in use and by 1940, nearly five hundred
thousand. In his *Office Duplicating*—which was printed in 1939
on an A. B. Dick Mimeograph machine—George H. Miller wrote,
"There is little doubt that stencil duplicating in America owes its
rapid and widespread growth to the mimeograph machines and sten-
cils as developed by the A. B. Dick Company."

Albert Blake Dick died on August 15, 1934, and his son, Albert
Jr., took over the business at that time. In 1949, the company relo-

cated to Niles, Illinois, a suburb of Chicago. By the mid-1970s, while the Xerox machine was rapidly replacing the mimeograph, the A. B. Dick Company had annual sales of $300 million and employed more than three thousand in the Chicago area.

As it declined, the firm was bought and sold by several concerns in the 1970s, '80s, and '90s. In 2004, the A. B. Dick Company filed for bankruptcy, and Presstek, a manufacturer of digital printing technologies, acquired its assets.

16

Linn Boyd Benton
1844–1932

Pantographic Punch-Cutting Engraver
New York, New York
1884

Linn Boyd Benton is not a widely known figure in the history of printing. This is an odd fact given that he is responsible for one of the most important technical achievements of the late nineteenth century: the invention of the pantographic engraver of type punches. Without Benton's contribution, the completion of the industrialization of the printing process—and the success of Mergenthaler's Linotype casting machine—would not have been possible.

Linn Boyd Benton was born on May 13, 1844, in Little Falls, New York, a town about seventy-five miles east of Syracuse. His father, Charles Swan Benton, was a lawyer and the founder-editor of the *Mohawk Courier* and *Little Falls Gazette*. In 1840, the elder Benton was elected as US Representative of the Seventeenth District of New York State.

It has been said that Linn Boyd was forced to rely upon himself at an early age because his mother, Emeline (Fuller) Boyd, of Little False died when he was just three years old. That his father moved the family frequently also contributed to Linn Boyd's character development. After Charles remarried, he relocated the family to Milwaukee,

Wisconsin, where he became part owner and editor of the *Milwaukee Daily News*. At age eleven, Linn Boyd had his first experience with typography in the composing room of his father's newspaper.

Boyd, as he was called, attended Galesville College in Galesville, Wisconsin, and studied advanced subjects for two years with a private tutor in La Crosse, Wisconsin. He developed his mechanical aptitude while working summer jobs as a tombstone cutter and as a watch repairman for a jeweler in La Crosse. At twenty-two years old, Boyd was hired by a friend of his fathers as a bookkeeper for a Milwaukee-type foundry. When the company went bankrupt during the financial panic of 1873, Boyd bought the Northwestern Type Foundry along with a partner and ran the manufacturing operations of the business. This was the beginning of Linn Boyd Benton's long career in typography.

After several name and partnership changes, Boyd became operations director of Benton, Waldo & Company, and by the early 1880s, the firm was manufacturing and selling metal type in the highly competitive industry. It was during this time that Benton began developing his skills as an inventor and typographic innovator. By the 1880s, the problem of standardized type-size measurement had become the scourge of the printing industry. Most printing establishments were forced to maintain relationships with a single type foundry since type sizes, widths, base alignment, and even metal alloy composition were not common.

With the industrial development of printing machinery long established—the launching of daily newspapers and installation of large steam-powered rotary web presses taking place everywhere—the lack of advanced methods of type specification, manufacture, and composition were holding the industry back.

Since the early 1700s, efforts had been mounted in Europe and America to come up with a standard for measuring type. The "pica and point" system finally emerged after a long conflict over proprietary interests. On September 17, 1886, the American System of Interchangeable Type Bodies was formally adopted at a meeting of the United States Type Founder's Association in Niagara, New York.

Within this environment, Linn Boyd Benton began working on methods that would change the way type was specified and handled. The problem facing compositors was that justifying a line of type required the manual arrangement of individual characters and spacers with a "trial and error" working method. According to Benton and others, this antiquated process unnecessarily lengthened composition time, and there had to be a means of automating it.

The measuring system of "12 points to a pica and 6 picas to an inch" that we use today was initially developed as a vertical system of type height. Benton's innovation was that the width should also be measured such that the typography followed "the point system both ways." In 1883, Benton received US Patent 290,201 for Self-Spacing Type that, according to the promotional literature, could "increase composition speeds by 25 percent."

The Benton, Waldo & Company's Self-Spacing Type styles were designed primarily for the newspaper industry where compositor speed was the most important issue. For some typographers, the horizontal distortion of characters and spaces required to make Benton's system work meant that the visual appearance of the type was unacceptable; they argued it was hard to read. Nonetheless, Benton's work on self-spacing type was a breakthrough, and the production and marketing of its typefaces brought him straight into another, much more historically significant, technological advancement for the industry.

To grasp the impact of Benton's invention of the pantographic punch-cutting engraver, it is important to understand the components and process of metal typesetting. Gutenberg's accomplishment was the invention of the handheld mold for typecasting. It created the mass production of individual metal characters that could be assembled into lines and pages of type, effectively displacing the handwriting by scribes. There are two preliminary steps required in the production of Gutenberg-type mold into which the molten metal is poured: the punch and the matrix. The punch is a steel relief form of the letter that is driven by a hammer into a piece of copper that creates the cavity of the matrix. The matrix is then placed into the mold assembly where hot metal is poured, forming the finished piece of type that will be inked and printed upon.

Prior to Benton, punch cutting was a manual process that required a highly skilled craftsman to design and engrave the characters into a tapered piece of steel that was two to three inches long. Every character of every size had to be punch cut; these were the "masters" of the font from which many matrices could be produced. If a punch was damaged or broken, it would have to be remade by hand, and it was likely that there would be slight differences from the one to the other.

Punch cutting was clearly the most difficult job in the production of type. It was not uncommon for a skilled punch cutter to take an entire day to make one punch; each punch required the continuous use of a magnifying glass, significant manual dexterity, and an aesthetic sensibility.

Benton's Self-Spacing Type required the cutting of more than three thousand punches, and skilled punch cutters were in short supply. To solve this problem, Linn Boyd Benton employed the pantograph—a mechanical device that uses parallelograms to trace an image on one surface and reproduce that image precisely on another surface—in the type production process.

Although Benton was not the first person to employ the pantographic principle in type making, he was the first to obtain a patent for the machine that would ultimately be used for cutting steel punches. The device went through several iterations, and it has been established that the first machine did not cut punches but was used to engrave the metal letters themselves. However, Benton's third pantographic punch-cutting engraver that was granted US Patent 332,990 was designed specifically for punch cutting.

Coincidentally, while Benton was solving problems with self-spacing type manufacturing, Ottmar Mergenthaler was developing a solution for the mechanical composition of type, one complete line at a time. With the investment of powerful newspaper publishing interests behind him, Mergenthaler invented the Linotype machine in 1886, and by 1888, there were hundreds of these machines on order.

However, Mergenthaler's Linotype breakthrough begged for a method of mass matrix production on a scale that had never existed. As explained by Benton's publicist, Henry Lewis Bullen, "Here was a machine, but no adequate means of supplying it with matrices had been devised. The rapid production of matrices required the rapid production of punches.… In 1890, the Linotype company had six or seven punch cutters in its employ, and these could do no more than keep up supply of matrices for about two hundred machines. Not in all the world could enough steel punch cutters be found to furnish an adequate supply of matrices without which the machines were as useless and unsalable as a gun where powder is unprocurable."

By chance, Benton's partner, R. V. Waldo, was on a self-spacing type sales visit at *New York Tribune* where Mergenthaler's machine was pioneered. Once the topic of Benton's pantograph came up between Waldo and Mergenthaler's representatives, it was just a matter of time before the Linotype matrix production dilemma would be solved. On February 13, 1889, the first Benton punch-cutting machine was leased to the Mergenthaler Printing Company. Thus, the combined accomplishments of Benton and Mergenthaler terminated the era of handcrafted-type production and enabled this most important aspect of print technology to completely enter the industrial age.

Linn Boyd Benton would go on to make many other technical contributions to the printing and typographic industries: combination fractions (1895), a type-dressing machine (1901), an automatic type caster (1907), and a lining device for engraving matrices of shaded letters (1913). Benton also played an important role with Theodore Lowe De Vinne in the design of the Century Roman typeface, an innovation in type design at the beginning of the twentieth century.

In 1892, Benton, Waldo & Company merged with twenty-three other type houses and formed the American Type Founders Company with headquarters in Elizabeth, New Jersey. At the time, it represented 85 percent of all type manufactured in the US and would dominate the industry into the 1940s. Morris Fuller Benton, Linn Boyd's only son, born in Milwaukee in 1872, would join the ATF organization at age twenty-four after graduating from Cornell with an engineering degree. Morris would go on to be a major contributor to the type business and a force of his own in printing history completing 221 typeface designs—including Cheltenham, Hobo, Broadway, and Franklin Gothic—during his career.

Linn Boyd Benton retired from ATF on July 1, 1932, and died two weeks later on July 15, 1932. Along with recognition for his many accomplishments, the company's board of directors described Benton in a statement the following October: "As a man, Mr. Benton endeared himself to us by his modesty, his delightful humor, and his probity in all matters, intellectual, and material."

17

Ottmar Mergenthaler
1854–1899

Linotype Machine
Baltimore, Maryland
1886

Ottmar Mergenthaler, the inventor of the Linotype machine, was born on May 11, 1854. While it may be difficult to appreciate in our era of digital innovation, Mergenthaler's invention was a momentous achievement in the history of print communications. Called the "eighth wonder of the world" by Thomas Edison, Mergenthaler's machine automated and integrated the casting and assembly of typeforms; the invention completed the advancement of printing from handcraft to industrial manufacturing.

In the early nineteenth century, industrial printing came into shape as iron replaced wood in press construction. Cylinders were developed, and with steam power, the rotary press was invented. Meanwhile, the 1800s saw papermaking industrialize and bookbinding convert from handiwork to mass production. The industrial revolution increased the volume of printing and its output per hour. However, in contrast, there was one critical step that remained primitive, costly, and in need of revolutionary change: typesetting.

For centuries, type composition methods remained essentially static. Standing in front of the typecase, the compositor manually picked up individually cast metal characters and placed them in sequence, including justification, onto an eight-inch composing stick. Manual typesetting required *The New York Times*, for example, to maintain a staff of one hundred compositors…to produce eight-page weekday editions and a twelve-pager on Sundays!

The search to replace the old process—which remained as it had been since Gutenberg's time—spanned a century and went through more than one hundred unsuccessful inventions. As one author wrote, "The story of the many inventors who failed, died broke, or took to drink…only serves to dramatize this persistent urge of the human spirit" to leap over an important technological hurdle.

With expectations of a solution on the horizon, investors put up an estimated $10 million between 1865 and 1885 for various schemes. Among them was Mark Twain who sank $190,000 ($7 million in today's money) into a failed contraption that had eighteen thousand moving parts. It took the genius, clarity of vision, and persistence of a young German emigrant to America to solve the riddle of mechanical typesetting. Arriving in Baltimore at the age of eighteen, Ottmar Mergenthaler was looking for an opportunity to pursue his interest in engineering.

In 1876, the young Mergenthaler—who was trained as a watchmaker—was asked to review the apparatus of another inventor. He said, "Even though I know almost nothing about printing, I have little faith that this is the machine to revolutionize an industry." And with that, he embarked upon the project of his life. After ten years of work and at age thirty-two, Mergenthaler demonstrated his machine publicly for the first time in the offices of *The New York Tribune* on July 3, 1886. With his unique arrangement of keyboard, gears, wheels, belts, pulleys, tubes, cables, hoses, trays, etc., Mergenthaler integrated a mind-boggling array of functions. At the core of his invention were (1) the casting of individual typeforms from hot metal, (2) the assembly of the characters into lines of type, and (3) everything was done by a single operator.

By 1895, there were three thousand machines in use by publishers and printers worldwide. The fast, low-cost typesetting led to an expansion of the printed word. In the US, newspapers increased pages, magazines grew in size and frequency, books were printed in greater quantities, libraries multiplied, and the illiteracy rate was reduced. Since he was a technician and not a moneyman, Mergenthaler was treated badly by the "publishers syndicate" that financed the development of the Linotype. Expecting rapid and large returns, they forced Ottmar to bring his invention to market before it was ready and manipulated the patent system for their own gain.

Along with other inventors of his generation, Ottmar Mergenthaler exhibited little interest in riches or personal glory. After twenty years of work and following a serious illness, he and his family took their first vacation. Traveling to his hometown in Wurttemberg, Ottmar was greeted with a hero's welcome. Upon their return to America, however, he was diagnosed with TB, and at the age of forty-five, Mergenthaler died on October 28, 1899.

18

William Morris
1834–1896

Kelmscott Press
London, England
1891

A polymath is a person with extraordinary expertise in multiple disciplines. Perhaps the most remarkable polymath of all time was Leonardo da Vinci, the Italian Renaissance painter, sculptor, inventor, engineer, musician, astronomer, anatomist, biologist, geologist, cartographer, physicist, and architect. In addition to painting the *Mona Lisa* and *The Last Supper*, da Vinci also conceptualized a flying machine, adding machine, solar power, and plate tectonics among many other innovations in the sixteenth century.

Any list of polymaths in world history would include Nicolaus Copernicus: astronomer, lawyer, physician, politician, and economist; Isaac Newton: mathematician, physicist, theologian, astronomer, and philosopher; Benjamin Franklin: author, printer, scientist, inventor, and statesman; and Thomas Jefferson: architect, paleontologist, inventor, horticulturalist, and politician.

It goes without saying that people with such a range of talents are extremely rare. Among the distinguishing attributes of a polymath are boundless curiosity and a facility for encyclopedic knowledge. Additionally, because they can acquire extensive practical experience in overlapping fields, polymaths are often responsible for inventions, discoveries, breakthroughs, and noteworthy creative works.

While it would be an exaggeration to place William Morris in the company of the abovementioned geniuses of human achievement, he was nonetheless a polymath. During his lifetime, William Morris made major contributions to architecture, textile design, decorative arts, poetry, literary fiction, politics, typography, and printing.

William Morris was born on March 24, 1834, in Walthamstow. He was the third child and oldest son of William Morris and Emma Morris Shelton. His father, who had moved to London from Worcester in the 1820s, became a partner in a stock brokerage firm in the city. His mother was the daughter of Joseph Shelton, a music teacher in Worcester.

In the early 1840s, the elder William Morris became very wealthy from an investment in a copper-mining business. The family then moved into a 150-acre estate in Woodford with its own brewery, bakery and buttery, and a governess and housekeeping staff.

Little William Morris was a precocious child and learned to read very early. By the age of four, it is said, he was reading books and familiarizing himself with the Waverly novels. At Woodford, William spent time exploring the outdoors, going fishing, and rabbit hunting, and he developed a lifelong appreciation for animal nature. After William Morris senior died at age fifty in 1847, the Morris family moved back to Walthamstow along with a considerable fortune.

One year later, William Morris was enrolled at Marlborough College where he attended for three years. In a recollection of his experience at Marlborough, Morris said it was "a new and very rough school. As far as my school instruction went, I think I may fairly say I learned next to nothing there for indeed next to nothing was being taught." Despite this negative memory, Morris spent time in the school library and developed an interest in archaeology and gothic architecture. He also took long walks in the countryside amid the Neolithic and Bronze Age monuments, stone circles, and burial mounds.

Following a rebellion and four-day strike in 1851 by students against the conditions at Marlborough, the Morris family removed

William from the college. Determined to prepare him for entry into Oxford, the family arranged a private classical tutor for the seventeen-year-old William. In 1852, William was admitted as a nonresident, and, unable to dine of sleep in the company of his fellow students, he entered Exeter College, Oxford.

It was at Exeter that William made the acquaintance of a student from Birmingham, Edward Burne-Jones, who would become a lifelong friend and collaborator. While the two students had entered the school with the intention of joining the priesthood, they both decided to dedicate themselves to the arts following a trip and tour of the great Gothic cathedrals in Northern France.

William then joined Edward—along with several other undergraduates and friends of Burne-Jones from Birmingham—in a group of intellectuals that called itself The Brotherhood. This club, which historians sometimes call the Birmingham Set, began meeting regularly to read theological tracts. This gave way to William Shakespeare, the poetry of Alfred Tennyson and Robert Browning, the novels of Charles Dickens, and then to a secular study of the art and architecture of the Middle Ages.

By 1855, the young men were influenced greatly by the views of John Ruskin and the Pre-Raphaelites who wanted to counter the influence of the industrial revolution upon artistic and cultural expression. Ruskin and his followers believed that art had to be returned to the handcraftsmanship that had been abandoned beginning with the works of Raphael. Ruskin taught that the separation of the intellectual work of the designer from the manual work of physical construction—a significant feature of mass industrial economy—was socially and esthetically damaging.

Returning to medieval artistic forms and techniques—and rebelling against what was considered the "barbarity" of contemporary industrial culture—would become a recurring theme of the subsequent works of William Morris. He developed the firm conviction that "without dignified, creative human occupation, people became disconnected from life." Alongside these cultural ideas, Morris and The Brotherhood advocated social reform aimed at improving the conditions of misery among the industrial workingmen of Victorian England.

In 1856, the members of The Brotherhood published twelve monthly issues of the *Oxford and Cambridge Magazine*—financed by Morris—which espoused the views of the group. It was out of these ideas that William Morris, Edward Burne-Jones, Charles Faulkner, Dante Gabriel Rossetti, and others would establish what later became known as the Arts and Crafts Movement that spread throughout the world in the late nineteenth and early twentieth centuries.

It is not possible to deal in detail with the many significant accomplishments made by William Morris in so many fields over the next four decades.

The following is a brief outline:

- Architecture

  He transformed domestic architecture and construction with the Red House built in Kent for Morris and his wife in 1959 with designs by Philip Webb. The house is made of red brick with a steep tiled roof, utilizing all natural materials. Morris founded the Society for the Protection of Ancient Buildings (SPAB) in 1877 dedicated to the repair and preservation of England's ancient buildings. The society still exists with 8,500 members and operates according to Morris's original manifesto.

- Decorative Arts

  Founded in 1861, the Morris, Marshall, Faulkner & Co. was established to create woodcarvings, stained glass, metalwork, paper hangings, and printed fabrics and carpets. To this day, the textile designs of William Morris—for furniture, embroidery, and wallpaper—remain among the most popular choices for home decor.

- Literature

  Morris wrote several novels—*News from Nowhere* (1890), *The Wood Beyond the World* (1894), and *The Well at the World's End* (1896)—that were the first works of science-fiction fantasy. Morris influenced both C. S. Lewis

(*The Chronicles of Narnia*) and J. R. R. Tolkien (*The Lord of the Rings*). In 1869, after learning Old Norse language and together with Eirikur Magnusson, Morris published translations of Icelandic mythology and folklore into English. The project eventually became a six-volume library of ninth-century Icelandic sagas.

- Politics

In 1883, Morris became an active member of the Social Democratic Federation and played a prominent role in its political work. Morris was the financier, editor, and writer from 1885–1890 for *Commonweal*, an influential English left political magazine.

Toward the end of his life, William Morris turned to the craft of printing and publishing. In 1891, he founded Kelmscott Press along with William Bowden near his home in Hammersmith, London. As with his previous artistic ventures, Morris wanted to shift book design and production back to that of medieval times. At Kelmscott Press, Morris set out to produce books with traditional methods as much as possible. This meant redesigning typefaces to reflect the look of the fifteenth century and simultaneously eschewing the use of lithographic printing systems. Morris believed strongly that contemporary book production was inferior to that which could be achieved by the craftsmen's handwork based on strict adherence to fifteenth-century techniques.

For all fifty-three books produced by Kelmscott Press, special handwoven paper was made entirely of linen, natural materials were used for custom-made inks, and Morris himself designed the typefaces. Based upon the 1470-type of Nicolas Jenson, Morris developed his Golden type in 1891. Later that year, he also designed the gothic Troy font based upon the black-letter type printed by Gutenberg in 1450.

Perhaps Morris's greatest printing accomplishment was *The Works of Geoffrey Chaucer* published by Kelmscott in 1896. The magnificent volume established a new standard for book design at the end of the nineteenth century. Adorned with eighty-seven illustrations and many decorative black borders of acanthus and vine—all designed and produced by Morris's friend, Burne-Jones—there were approximately 425 copies printed.

Morris's contribution to book design was summed up in a talk that he delivered to the Bibliographic Society in 1893 called "The Ideal Book." In his presentation, Morris exhibited his considerable knowledge of the technology of book printing as well as the esthetics of typographic design. In poetic style, Morris said, "First, the pages must be clear and easy to read, which they can hardly be. Secondly, the type is well-designed, and thirdly, whether the margins be small or big, they must be in due proportion to the page of letter."

William Morris—the polymath designer, author, social theorist, and printer—died on October 3, 1896, at the age of sixty-two. A major figure of the late nineteenth century and during a time of great technological change, Morris sought solutions to the dilemmas of his time in medieval styles and methods. While one may rightfully question his romanticism and singular identification of industrial progress with "barbarity," the positive influence of William Morris lives on today in many more ways than is popularly appreciated or understood.

19

Karel Klíc

1841–1946

Rotogravure Printing
Vienna, Austria
1895

Most people are familiar with *National Geographic* magazine. Published monthly and since 1888, *National Geographic* has a worldwide circulation of more than eight million copies and appears in thirty-six languages. Widely known for the yellow rectangular border of its cover and the use of dramatic color photographs of world geography, history, and culture, it is one of the most popular magazines ever printed.

The name "National Geographic" is used occasionally as a printing industry euphemism. Someone might say "It's not *National Geographic*" when speaking of color reproduction expectations of less demanding projects. What most are unaware of, however, is that *National Geographic* achieves such beautiful and spectacular color largely because of the gravure method used in the printing of its editorial pages.

Although less common in publication printing than offset lithography, gravure printing is typically associated with magazines that have very large circulation. According to Hans Wegner, VP of Production Services at *National Geographic*, gravure printing provides superior color saturation and consistency, a more photographic look as well as important cost advantages.

The gravure process—a form of intaglio printing—involves engraving an image carrier, typically a metal cylinder, with recessed cells. The cylinder is immersed and rotates in fluid ink. As the cylinder turns, ink fills the imaging cells, and before contacting the paper, a doctor blade scrapes the excess ink off the cylinder in the nonimage area. The paper is brought into contact with the inked cylinder by an impression roller, and the ink is drawn out of the cells onto the paper by capillary action.

The high-quality reproduction of gravure printing is the result of the following: very fine halftone dot sizes that emulate the grain of continuous tone photography, greater image detail than offset printing and a CMYK color gamut than that is often wider than that of offset printing because a greater amount of ink pigment is transferred the surface of the paper.

The history of gravure printing is complex and poorly documented. Dealing with the lack of reliable historical information in his *History of Industrial Gravure Printing* up to 1920, Otto M. Lilien wrote, "More and more of the technical development is described with only sketchy details, and it is noticeable that the references are often missing. Frequently, the information contradicts itself regarding the person credited with inventions and technical improvements."

The manual-gravure printing process was created and perfected in the nineteenth century, and the earliest inventions are associated with photography. In 1826, Joseph-Nicéphore Niépce developed the first photomechanically etched printing plate that was made of zinc and used to print portraits. In 1852, William Henry Fox Talbot developed a method for making gravure printing plates that could transform a continuous tone picture into a halftone.

According to Lilien, Paris publisher Auguste Godchaux took out the first patent for a gravure-printing press that used cylinders and printed on a web of paper in France in 1860. Godchaux built the press, and it ran for eighty years in a printing facility in Paris on Boulevard Charonne until the Nazi's occupied the city in 1940.

Karel Klíc (Karl Klietsch) is the recognized inventor of modern gravure printing. Although it is sometime stated that Klíc developed all the complex gravure processes without knowledge of the work of others, he was the first to bring all of them together. According to Lilien, Klíc integrated the crossline screen and the transfer of gelatine pictures to metal plates for cylinder production.

Karel Vaclav Klíc was born on May 31, 1841, in Hostinne, in the foothills of the Krkonose Mountains in the present-day Czech Republic and about thirty-five kilometers from the border with Poland. The town has been known as a center of papermaking.

Klíc showed interest in the arts and at the age of fourteen was admitted to the Art Academy in Prague where he was expelled for nonconformance in 1855. He later returned to complete his studies. As a young man, Klíc worked as a draughtsman, a painter, illustrator, and cartoonist. With his latter skills, he worked at newspapers in Prague, Brno (Moravia), and Budapest before opening a photographic studio in Vienna, Austria, in 1883.

While in Austria, Klíc joined the Photographic Society of Vienna and was introduced to many of the new developments in photo-reproduction methods. His early attempts at photogravure techniques were presented with much acclaim at the annual society exhibitions in 1879 and 1880. During these years, Klíc did not reveal anything publicly about his methods. Recognizing the monetary value of the process he had perfected, Klíc sold the process to others in Vienna and London.

In 1880 and 1881, several of Klíc's photos were published in an Austrian journal, *Photography Correspondence*. In 1882, a heliogravure portrait of Mungo Ponton—a Scottish pioneer in photographic techniques and an amateur scientist—was reproduced as a special insert to the British *The Yearbook of Photography and Photographic News Almanac*.

Writing about the significance of the image of Ponton, the editor of the almanac wrote, "We ought to say a word about our portrait of Mungo-Ponton, an Englishman who may well be termed the discoverer of permanent photographic printing, for it was he who proposed, in 1839, the employment of bichromate in photography.

Klíc's is an etching process upon copper, an imprint from a carbon diapositive being secured upon that metal. The mode of preparing the copper is a secret, but we may mention that the process is so quick that within four or five days, an engraved plate may be produced of considerable dimensions. Of the quality of the printing, our readers can judge for themselves. Suffice it to say, the process is an inexpensive one, and that during the past year alone, no less than three hundred photo engravings were produced."

After one of his business associates by the name of Leonard published the details of his process in an Austrian technical journal in March 1886, Klíc left the country in frustration and traveled to England. It was during this trip that Klíc met Samuel Fawcett, a process worker at Storey Brothers, a calico-printing firm located in Lancaster.

It is known that Klíc's vision for gravure reproduction extended beyond single-sheet photographic prints. Fawcett had been working independently in 1890 on a series of gravure experiments, and his contact with Klíc was the catalyst for the development of an entirely new industrial printing system. Klíc and Fawcett, beginning in 1895 with the formation of the Rembrandt Intaglio Printing Company, jointly developed the rotogravure process—modern gravure printing. The men experimented with screens of 150 and 175 lines per inch and printed on paper with machines owned by Storey Brothers and designed for printing on textiles.

The process developed by the Rembrandt Intaglio Printing Company remained secret for ten years, giving the firm a lucrative monopoly on the process before any competitors emerged in the market. In 1897, while technical director of the company, Klíc left England and returned to Vienna to continue with further experimentation and invention. He came back for a short time in 1906 after he perfected a method for three-color gravure process with fine halftone screens. Karel Klíc died in Vienna on November 16, 1926.

20

Frederic Goudy

1865–1947

Metal Type Design
Park Ridge, Illinois
1903

n reviewing the life of Frederic Goudy—the most important American type designer in the first half of the twentieth century—we see a man who thrived during an era of terrific change. Born at the conclusion of the Civil War, Goudy lived through the industrialization of society and two world wars. Without a formal education, self-taught, and often under difficult circumstances, Goudy became a prolific type designer who was known the world over for his accomplishments. Significantly, Frederic Goudy drew his first alphabet at age thirty and began his career as a professional type designer at the age of forty-six.

Frederic William Goudy was born on March 8, 1865, about 125 miles south of Chicago in a town called Bloomington, Illinois. As a youngster, Fred spent time in the Bloomington library reading Mark Twain and browsing the illustrated *Harper's Weekly* magazine. He developed the ability to trace and replicate wood engravings in pencil. Although he did not excel at mathematics, Fred gained an interest in machines such as the lathe and the pantograph.

Fred's father, John Goudy, was a school administrator and later a real estate man. He moved the family to four different towns in

Illinois in the 1870s and early 1880s. In 1883, the family relocated to the Dakota Territory where John started a business in connection with homestead claims. Living in the prairie hamlet of Highmore near an Indian reservation, and with just two years of high school education behind him, Fred went to work as a clerk in his father's real-estate company.

In 1889, Frederic left Highmore and set out on his own. First going to Minneapolis and then Springfield, Illinois, he worked as a bookkeeper in real-estate offices. During these years, Fred gained experience with advertising and layout of newspaper ads. At twenty-eight years old, Fred moved to Chicago and worked for various offices writing advertising copy and designing ads with local printers.

In the 1890s, the advertising industry barely existed, and there were a handful of advertising agencies. Newspapers and magazines were filled with garish promotional ads with very bad typography and florid graphics. At this time, Fred started swimming against the tide of generally murky and unreadable printing. In 1893, he founded a magazine called *Modern Advertising* as a means of generating business. Although the publication did not last, Fred gained important experience with the typography and print-production processes.

Around 1895, Goudy became influenced by the works of William Morris—the English poet, author, craftsmen, and designer—and the Arts and Crafts Movement. In collaboration with a Chicago English teacher C. Lauron Hooper, Goudy decided to start his own printing business along the lines of William Morris's Kelmscott Press in England. He would later say of this time, "When I became inoculated with printers' ink, I was never again the same."

As an amateur, Goudy began experimenting with type designs and developed his hand lettering skills. After a short time in Detroit working for a weekly called *The Michigan Farmer*, Goudy returned to Chicago and worked on advertising for *Marshall Field*, *The Inland Printer*, *The Pabst Brewery*, and *Hart Schaffner and Marx*. He also designed book covers for The Lakeside Press and Rand-McNally. All Goudy's type designs through this period were for advertising purposes.

By the turn of the century, Goudy wanted to follow Morris's lead and print the finest books in America. To do so, he believed he needed to design his own typeface. In 1903, Frederic Goudy made printing history with the establishment of the Village Press in Park Ridge, Illinois, and the creation of the Village Type, his first fine book face. The Village Type was the very first American typeface to be cut and caste from free hand, original drawings from a type designer.

In 1904, Frederic and his wife, Bertha, moved the Village Press to Hingham, Massachusetts, to become part of the Hingham Society of Arts and Crafts and be surrounded by other craftsmen. In 1906, the Goudy's moved their printing business to New York City. It was during a trip to England in 1909 and then a trip to the European continent in 1910 that Goudy focused himself upon the scholarship and history of typography.

In 1911, according to his own account, Frederic Goudy became a professional type designer with the creation of Kennerley Old Style. Named for his business associate, publisher and Englishman Mitchell Kennerley, Goudy designed the font specifically for the publication of H. G. Wells's *The Door in the Wall and Other Stories*. Kennerley Old Style was a hit in England, and through it, Goudy suddenly became associated internationally with great type design. It would take some years more for his work to become recognized in America.

From this point forward, Frederic Goudy devoted his energies to type design, and his hand lettering and printing work receded into the background. Over five decades, Goudy designed 122 typefaces, an enormous accomplishment in the era of hot-metal typography.

Among Goudy's popular typefaces are

- Copperplate Gothic (1905),
- Goudy Old Style (1915),
- Hadriano (1918),
- Italian Old Style (1924),
- Trajan (1930), and
- Berkeley Old Style (1938).

A significant technology factor in the emergence of type design as a profession—making it possible for someone like Frederic Goudy to achieve success—was the invention by Linn Boyd Benton in 1884 of the pantographic punch-cutting engraver. This device, which represented the industrialization of metal-type production, enabled foundries to cut matrices from enlarged drawings. Prior to this development, the making of type was largely the work of handicraft punch cutters and not that of designers. At the age of sixty, Goudy acquired his own matrix-cutting machine on which he engraved and cast perhaps some of his greatest work.

During his career, Goudy wrote extensively on type design, lettering, typographic style, and history. His works *The Alphabet* (1918) and *The Elements of Lettering* (1922) remain important resources, the latter containing explanatory notes on the considerations and influences behind some of his typeface designs. He founded the journal *Ars Typograhica* in 1918, and he became the art director for Lanston Monotype Corporation in 1920 where he remained until his death.

Frederic Goudy was famous during his lifetime. He was a rugged man, a widely read commentator on design and aesthetics, and a popular speaker who was approached for his opinion on many topics, some far from his field of expertise. He was known for his larger-than-life personality, as a raconteur, and he could be counted on for comments with a punch line. It is said that Goudy was the originator of the statement "Anyone who would letterspace black letter would shag sheep" although he may have used a different word than "shag."

For some of his competitors, Goudy's self-promotion was a problem. Historically, type designers had never named their works after themselves; Goudy used his name in about twenty of his typefaces. Goudy's love for typography is summed up in his favorite broadside designed and produced for a Chicago exhibition in 1933:

> I Am Type! Of my earliest ancestry neither history nor relics remain. The wedge-shaped symbols impressed in plastic clay by Babylonian builders in the dim past foreshadowed me from them, on through the hieroglyphs of the ancient

Egyptians, down to the beautiful manuscript letters of the mediaeval scribes, I was in the making.

With the golden vision of the ingenious Gutenberg, who first applied the principle of casting me in metal, the profound art of printing with movable types was born. Cold, rigid, and implacable I may be, yet the first impress of my face brought the Divine Word to countless thousands.

I bring into the light of day the precious stores of knowledge and wisdom long hidden in the grave of ignorance. I coin for you the enchanting tale, the philosopher's moralizing and the poet's fantasies. I enable you to exchange the irksome hours that come, at times, to everyone for sweet and happy hours with books—golden urns filled with all the manna of the past. In books, I present to you a portion of the eternal mind caught in its progress through the world, stamped in an instant, and preserved for eternity. Through me, Socrates, and Plato, Chaucer and Bards become your faithful friends who ever surround you and minister to you.

I am the leaden army that conquers the world. I am type!

Frederic W. Goudy died at his home in Marlboro-on-Hudson, New York, on May 11, 1947. He is buried next to his wife, Bertha, in Evergreen Cemetery in Chicago.

21

Ira W. Rubel

1860–1908

Offset Printing
Nutley, New Jersey
1904

Walter E. Soderstrom, the noted authority on print technology, wrote in the *Photo-Lithographer's Manual* of 1937, "The origin of the offset press is one of the least discussed subjects in the literature of printing." In preparing this brief sketch of the life of Ira Rubel and his invention of the offset printing press—an extraordinarily important event in the history of modern print technology—it is evident that nothing much has changed since Soderstrom's time. To this day, there is little accessible and authoritative information about Ira Washington Rubel: his work or his life. In fact, most of what is easily found either trivializes Rubel's contribution—as purely happenstance rather than the work of an earnest inventor—or glosses over its historic significance. There is currently no biographical Wikipedia entry on the man and his accomplishment; photos of Ira Rubel are very hard to find.

Ira Washington Rubel was born in Chicago on August 27, 1860. He attended Hayes and Division West High Schools in Chicago. He graduated from the University of Chicago in 1881. Rubel then attended Northwestern University in Evanston where he was a classmate of William Jennings Bryan and graduated with a degree in

Bachelor of Law in 1883. He litigated cases as a practicing attorney for a short time in Chicago.

The printing firm, Rubel Brothers, which Ira founded along with his brother Charles, is listed in the Lakeside Annual Directory of the City of Chicago of 1887. During the 1880s, Ira became known for having been a pioneer in the manufacture of loose-leaf systems. Promoting the systems for the maintenance of business transactions records, the Rubel Loose Leaf Mfg. Co. was established on Superior Street in Chicago.

By 1899, the Rubel Brothers' firm acquired both a six-story building on Wabash Street in Chicago as well as a paper mill. At some point, Ira Rubel and his three brothers expanded to the east coast and established an office on Broadway in New York City. They also opened a lithographic printing and paper mill facility in Nutley, New Jersey. By 1901, the Rubel Brothers Paper & Manufacturing Co. on Kingsland Street in Nutley, experienced significant growth and expanded its facilities. It was at this plant that Ira would conduct his experiments, discover the offset printing method, and build the first offset lithographic printing press.

Offset printing is also known as "indirect" printing, that is, the printed image is not applied directly to paper from the printing plate or inked image carrier but is first transferred to a rubber blanket and then to paper. Although indirect ink transfer had already proven its importance for at least twenty-five years in tinplate (canned goods) printing, the concept was not obvious in paper-based lithographic printing due to the dominance at the time of the letterpress technique, especially for typographic reproduction.

Perhaps one of the reasons for the lack of literature on Rubel's discovery is the fact that the offset method was the result of more than fifty years of seemingly disconnected technological developments in both typography and pictorial reproduction. In 1875, Robert Barclay of Barclay & Fry obtained an English patent for the very first offset press that transferred the printed image from the lithographic stone to a cardboard surface and then to sheet metal (used in making biscuit tins). The discovery of halftone photographic and process-color reproduction in the 1880s was also a factor that motivated the invention of offset.

In 1903 and 1904, Rubel experimented at the Nutley facility with photo reproductions transferred onto lithographic stone through a screen. Although the testing did not yield significant results, Rubel's work on a stop-cylinder press led inadvertently to an important breakthrough. When his assistant misfed a sheet and the rubber impression roller came into contact with the lithographic stone, the reversed image was transferred to this roller. When the next sheet was fed, it had an image on both sides: one that was the product of direct contact with the stone image carrier as intended and the other with a wrong-reading image from the "indirect" rubber roller.

Once the "indirect" image was found to be superior in quality to the direct one, Rubel and his collaborators expanded their testing along these lines and perfected the technique. This included a complete redesign of the press based upon the offset principle.

As Rubel was convinced of the importance of his discovery, he returned to Chicago to seek the technical assistance of lithographer Alexander B. Sherwood. The two also enlisted the financial support of Andrew Kellogg of New York as a venture partner. Although the arrangement did not last, three presses were built, and each man took one of them as their share of the achievement.

Court rulings made it impossible for Rubel to obtain a US patent for his discovery because the preexisting use of the offset method by tinplate printers was legally invoked. Therefore, each of the partners—Rubel, Sherwood, and Kellogg—went their separate ways with the invention. The Kellogg Offset Press was built in New Hampshire; Sherwood joined the Potter Printing Press Company of Plainfield, New Jersey; and Ira Rubel took his idea to England where he thought he might obtain a patent for it. In 1906, Rubel met the New Zealander Frederick Sears, and together, they launched what became known as the Sears "Highlight" process.

Rubel and his family never realized any fortune from his invention. He died of a stroke at the age of forty-eight while at the Derby Hotel, Bury, Lancashire, according to relatives, from "the worry and work occasioned in seeking to protect his patents and marketing his inventions in Europe and America."

An obituary published in the September 19, 1908, edition of *The American Stationer* reported, "Ira W. Rubel, pioneer manufacturer of loose-leaf systems and inventor of the offset printing press which did much in the progress of lithographic development, died suddenly in London of apoplexy. Mr. Rubel was formerly connected with the Rubel Brothers' Company, which was succeeded by the Rubel Manufacturing Company. Mr. Rubel left Chicago ten years ago and took up his residence in New York. His remains will be buried in Chicago in Graceland Cemetery by the side of his wife."

There is no doubt that Ira Rubel's invention—and the independent and concurrent work of Charles and Alfred Harris of Niles, Illinois—came to transform the printing industry worldwide. The offset lithographic process was greeted initially within the industry with mixtures of enthusiasm, skepticism, and opposition. As with many previous and future breakthroughs in the graphic arts, there were those who could see offset's potential and those who were advocates of the dominant letterpress of the previous technology generation.

It would take fifty years before the superior quality, speed, and economics of offset overtook letterpress as the dominant printing method internationally. By 1960, most printed matter was being produced on offset equipment. Today, with digital printing systems advancing rapidly in recent decades, offset still represents more than half of worldwide annual printing shipments. More than a century after Ira Rubel's invention, the dominance of offset printing will continue for many years—some even say decades—to come.

22

Stanley Morison
1889–1967

Times New Roman Typeface
London, England
1930

In 1930, Stanley Morison wrote, "Typography is the efficient means to an essentially utilitarian and only accidentally aesthetic end for the enjoyment of patterns is rarely the reader's chief aim. Therefore, any disposition of printing material which, whatever the intention, has the effect of coming between the author and the reader is wrong."

This is taken from Morison's essay *First Principles of Typography*, which became for decades an industry manual of book typesetting standards, especially in America. By the 1930s, Stanley Morison had acquired a remarkable depth of knowledge and experience in printing. He understood better than most the importance of the "invisible" beauty and subordination of form to function in typography.

Morison's first principle still applies today: when it comes to typographic design and style, especially in books, easing the comprehension of the text is the primary objective. "Dullness and monotony" and "obedience to convention" are preferred over "eccentricity or pleasantry" and "typographical experiment." A text is useless if it is hard to read because it is "different" or "jolly."

One might conclude from the above that Stanley Morison was opposed to typographic innovation, but nothing is further from the truth. Within a year of writing his tribute to typographical tradition, Morison would develop and design—in collaboration with the graphic artist Victor Lardent—one of the most widely used typefaces in history: Times New Roman.

Stanley Arthur Morison was born May 6, 1889, in Wanstead in Essex between London and Epping Forest. As a child of seven or eight, the family moved to north London. Stanley lived at this London location on Fairfax Road, Harringay, until he was twenty-three years old.

Stanley was largely self-taught. He left school at age fourteen to find work after his father—who was a traveling salesman—abandoned the family. His mother was strong-willed and inspired Stanley to serious and independent study. He was influenced by her to take up philosophy and a study of ancient manuscripts (paleography), spending his spare time at King's Library at the British Museum.

At age twenty-three, while working unhappily as a bank clerk, Stanley read a supplement published by *The Times* that carried an ad about the start of a new magazine on printing called *The Imprint*. After the first issue appeared in January 1913, he applied for and was hired as an editorial assistant. This job proved to be the start of the extraordinary graphic arts career of Stanley Morison.

During World War I, Morison was imprisoned for being a conscientious objector. Following the war, Stanley underwent a conversion to Catholicism and began a study of liturgical writings, hymnals, and other early church publications. In 1919, he became design supervisor for Pelican Press and in 1921 published his first typographical study: "The Craft of Printing: Notes on the History of Type Forms."

In 1922, along with Francis Meynell, Holbrook Jackson, Bernard Newdigate, and Oliver Simon, Stanley became a founding member of the type-centric Fleuron Society (a fleuron is a typographer's floral ornament). Seven volumes of their journal called *The Fleuron* appeared between 1923 and 1930. Each lavish edition contained papers, illustrations, specimens, and essays by contemporary authorities on typography and book design.

The scholarly works contained in *The Fleuron* remain relevant today. The theoretical ideas and concepts developed by the Fleuron Society are applicable to modern online and digital media as well as print. It was in this publication that Morison's *First Principles of Typography* originally appeared.

In 1923, Stanley Morison became a typographic consultant for the Monotype Corporation. Monotype was a manufacturer of hot metal-casting machines that industrialized and revolutionized in the 1880s—along with Mergenthaler's Linotype Corporation—the process of making type for print. While the Linotype machine cast complete lines of type primarily for newspaper publishing, the Monotype machine cast individual characters and was often used in book and other "fine" printing.

During the remainder of the 1920s, Morison became involved in Monotype's program of old-style type revival. Sparked by technological innovation, the first few decades of the twentieth century witnessed a typographic renaissance as Monotype fonts, such as Bodoni, Baskerville, Bembo, Centaur, Perpetua, and others, were reinterpreted and recut under Morison's direction.

In 1929, Stanley Morison publicly criticized *The Times* for being poorly printed and typographically antiquated. Following discussions with the publisher, Morison was hired as a consultant and commissioned in 1931 to develop a new, easy-to-read typeface for the newspaper. His task was to design a font that was economical—capable of fitting more copy in a column than previous typefaces—as well as technically compatible with the printing machinery of the time.

Morison began his work with an authoritative historical survey called "The Typography of *the Times*" that showed the evolution of its typography. Morison presented to the publisher a folio with forty-two full-size reproductions from the earliest days in the eighteenth century into the 1920s and made the case for his solution.

In *A Tally of Types* in 1953, Morison wrote that he "penciled the original set of drawings and handed them to Victor Lardent, a draughtsman in the publicity department of Printing House Square," who Stanley "considered capable of producing an unusually firm and lean line." The drawings were then used by Monotype to cut the punches for the first set of Times New Roman types. The first issue of *The Times* to use the new typeface appeared on October 3, 1932.

As a historian, Morison was appreciative of the accomplishments of others before him that made his work possible. In summing up the experience with the typography of *The Times*, Morison explained, "Above 14,750 punches, including those corrected (a large number), were cut by Monotype Corporation for the installation at Printing House Square.... Their cutting was a triumph for the mechanism invented by Linn Boyd Benton of Milwaukee. In 1885, he adapted the pantograph principle to the mechanical cutting of the punches used for striking the matrices from which the type is cast. This invention lies at the basis of all mechanical composition, which requires at some stage the pouring of metal into a single matrix or line of matrices."

Following the achievement of Times New Roman, Stanley Morison continued his design consulting work with Monotype and *The Times* for three decades. He became editor of the "History of the Times" from 1935 to 1952, and he was also editor of "The Times Literary Supplement" between 1945 and 1948. He spent his later years on typographical research. Although he was offered a knighthood in 1953 and the Commander of the Most Excellent Order of the British Empire (CBE) in 1962, he declined both. He was elected a Royal Designer for Industry in 1960. He died on October 11, 1967, at the age of seventy-eight.

It is difficult to appropriately summarize the work of a figure such as Stanley Morison in this short sketch. Although his writings have never been brought together into a single collection or set of volumes, Morison was a prolific scholar and practitioner of the graphic arts. He was perhaps the most important theoretician, designer, and historian of print of the mid-twentieth century.

Postscript

In 1994, printing historian Mike Parker published findings that showed Times New Roman was based upon a design originally made by William Starling Burgess in 1904. A complete review of Parker's story can be found in an article titled "The History of the Times New Roman Typeface" on the *FT Magazine* website. Although still controversial, *The Times* began in 2007 accepting the possibility of an alternative history to the one provided by Morison about the origin of the famous font. According to *The Times* website, Times New Roman was design by Morison, Lardent, "and possibly Starling Burgess." In 2009, Mike Parker worked with Font Bureau, Inc. and published a font series called Starling based upon Burgess's original conception.

23

Chester Carlson
1906–1968

Xerography
Astoria, Queens, New York
1938

Chester Carlson, the inventor of xerography, was born on February 8, 1906, in Seattle, Washington. His early years were filled with hardship. His family was poor, and his father suffered from multiple illnesses. Chester began working to support his family at the age of eight. When his mother died of TB, Chester was just seventeen. He would later say, "That is the worst thing that ever happened to me. I so wanted to be able to give her a few things in life."

Chester developed an early interest in printing. He started a newspaper called *This and That* at the age of ten and circulated it among his friends. He used a Simplex Typewriter to set type one character at a time. He said of this experience, "I was impressed with the tremendous amount of labor involved with getting something into print…and I got to thinking about duplicating methods."

Chester excelled in math and science and was encouraged by teachers to continue after high school. He attended Riverside Junior College and then Caltech, graduating with a degree in Bachelor of Science in Physics in 1930. Chester took his first job with Bell Labs in New York City as a research engineer. He would later transfer to the patent department as an assistant and turned his attention to document management. Chester recalled, "The need for a quick, satisfactory copying machine that could be used right in the office seemed very apparent to me… So I set out to think of how one could be made."

Chester's work on an office copier began in the mid-1930s. He conducted experiments with the help of Austrian physicist Otto Kornei, and their first breakthrough was achieved on October 22, 1938. They successfully transferred an image from a microscope slide to a sheet of wax paper using an electrostatic charge and some organic powder. Initially calling the process electron photography, Chester later commented, "The powder image was adhering to the plate by virtue of relatively small but nevertheless real, electrostatic forces."

With the help of a law degree obtained from the New York Law School in 1939, Chester successfully patented electrophotography in 1942. He tried to sell the concept to companies he thought might be interested in its commercial development. He wrote to more than twenty companies—including GE, IBM, A. B. Dick, and RCA—none of which took him up. He described their response as "an enthusiastic lack of interest."

In 1946, with the assistance of Battelle Memorial Institute of Columbus, Ohio, Chester finally convinced researchers and executives at the Haloid Company of Rochester, New York, to sign a $10,000 contract to license electrophotography. Marketing concerns turned Haloid to search for a better product name, and xerography was suggested, the combination of the Greek words *xeros* (dry) and *graphein* (writing).

The Haloid Company brought the Xerox Model A Copier to market in 1949, eleven years after Chester Carlson's discovery. However, it was not a commercial success. It would take another eleven years (and many technological developments) before the fully automated Xerox 914 would become a huge hit as the first plain paper-office copier. By 1962, ten thousand units had been sold, and by 1968, revenues for Haloid Xerox had reached $500 million.

By 1965, Chester Carlson was worth several hundred million dollars from royalties on his patents, making him one of wealthiest people in America. However, Chester spent years quietly giving away most of his fortune to charities. He died of a heart attack at the age of sixty-two on September 19, 1968.

Chester Carlson's invention—which took two decades to convert into a viable product—is used today in tens of millions of photocopying machines and laser printers, and it is the basis for digital printing systems that use toner, such as the Xerox iGen press. Along with digital ink-jet printing devices, xerographic systems are slowly moving in on traditional offset lithography as the dominant technologies of the printing industry.

24

John Crosfield

1915–2012

Printing Press Automation
London, England
1947

Today's digital and mobile wireless technologies are in a constant state of flux. By the middle of the second decade of the twenty-first century, the human computer interface was once again transformed with haptic technology—tactile feedback from a device such as force or vibration. If you have felt vibration in response to a touch function on your smartphone, then you have experienced haptics. What was until recently available only to virtual reality enthusiasts and gamers is now a feature of every smartphone and tablet.

Challenged to keep up with the pace of change, it is easy to miss the fact that the electronics revolution has been underway for more than a century, and digital electronics represents less than half that period of time.

Electronic technology can be divided into two basic forms: analog and digital. Long before there were microprocessors and memory chips that exchange all information, data, code, signals, etc. in a series of zeroes and ones, there were analog electronics, such as resistors, capacitors, inductors, diodes, and transistors. The difference between a clock with hour, minute, and second hands rotating around the face and the numerals on a Light Emitting Diode (LED) clock display is a simple illustration of analog vs. digital technology.

John Crosfield's contributions to printing and the graphic arts spanned both analog and digital electronics. His analog systems were developed in the late 1940s and became dominant in the industry throughout the 1950s. When the first computers were introduced in the 1960s, Crosfield pioneered digital electronics in the graphic arts and became a major worldwide provider of equipment into the 1960s and 1970s.

John Fothergill Crosfield was born into a well-off family. He was the third child and second son of prominent English Quakers. Born on October 22, 1915, in Hampstead, London—a community known for its intellectual, liberal, artistic, musical, and literary associations—John had five siblings.

John's father, Bertram Fothergill Crosfield, was managing director and coproprietor of the *News Chronical* and *The Star*, both liberal daily newspapers in London. Bertram was also leader of several Hampstead organizations. John's mother, Eleanor Cadbury, was the daughter of the famous chocolate maker and leading Quaker, George Cadbury. Eleanor was well-known independently of her father and was elected as a Liberal to Bucks County Council.

John showed an early interest in building things. As a boy, he was often busy in the family workshop making boats, steam engines, and other mechanical devices. He once built a cannon and tested it on the garage door. The projectile went through the door and damaged his father's Daimler. He was fond of trains and, with the assistance of a childhood friend, built an O-gauge model railroad on the property of his school grounds.

At age thirteen, John was enrolled in Leighton Park School, a Quaker establishment. He enjoyed studying physics and math and decided he wanted to pursue engineering at college. He enrolled at Trinity College Cambridge, following in his father's footsteps. He designed and built gliders and other flying machinery, such as a winch launcher, in his spare time. Although he had many hobbies, John was an exceptional student and put most of his time into his studies.

John graduated from Cambridge in 1936 and went to Munich, Germany to improve his language skills. He witnessed anti-Semitism and Nazi propaganda first-hand and was horrified by Hitler's methods. Upon his return to England, John's accounts of the treatment of political prisoners in Germany were met with disbelief.

John Crosfield was a member of a generation of engineers whose formative experiences were made in World War II. Much of the technology advancements that were deployed throughout industry in the postwar period originated in the struggle between the warring countries for military supremacy.

After he left Cambridge, Crosfield took a student-apprentice engineering position with British Thomson-Hudson (BTH), a heavy-industry firm based in Warwickshire. BTH was founded as a subsidiary of the US-based General Electric Company (GE) and specialized in steam turbines. In 1938, he left BTH and went to work at the Stockholm facility of ASEA, a Swedish version of BTH and GE. When the war began in 1939, John made his way back to England and planned to join the Navy.

Crosfield used some connections at ASEA to get an assignment by the admiralty to the Mine Design Department. It was here that Crosfield's electronic genius would begin to be expressed. He worked on a magnetic mine project that could detect German boats near British harbors.

Crosfield also designed and built a prototype of an acoustic mine that could pick up on the sound of the propeller of wooden German E-boats. The acoustic mine became a success with two hundred being deployed in the Baltic Sea and sinking forty-seven enemy vessels. Crosfield and his colleagues later worked on the development of both acoustic and subsonic mines. He got involved in the production process, and in 1944, Crosfield's inventions proved extremely effective in major battles at the Straits of Dover and the Western Approaches.

After the war, John Crosfield decided—learning from his experience at ASEA that some of the projects that he had worked on would never be funded—to start his own business. In 1947, he set up a lab in Hampstead and began working on new projects. He later recalled that in 1945, while he oversaw electronics research for the admiralty, he was approached by a printing-industry representative about the problem of color registration on high-speed rotogravure-magazine presses. There was a need for an automated system to align all the process colors in the printed page to improve quality and reduce press waste.

With about £2,000 of his own money and another £2,500 borrowed from family members, Crosfield set out to design an electronic and automated registration system for color printing. After eighteen months of hard work, the Autotron was tested as a prototype on the production of *Women's Weekly* at Amalgamated Press in London. Prior to the Autotron, a magazine production run would often waste 25–30 percent of the impressions using manual controls. Crosfield's automatic register system brought the waste figures down to 4–5 percent.

The Autotron consisted of scanning heads mounted on each printing unit and a control cubicle that was located away from the press. The scanning heads picked up register marks—unobtrusive symbols on the printed page that were hidden from view—to regulate the movement of the printed image from unit to unit with an accuracy of a one thousandth of an inch.

Word about the Autotron traveled quickly in the printing industry, and Crosfield was soon taking prepaid orders from companies in Britain. An opportunity to show the system at British Industries Fair in 1949 made Autotron an international phenomenon, and orders were quickly being placed from printers in countries around the world.

The success of the Autotron encouraged John Crosfield to invest in further research in pressroom automation for gravure-magazine printing and other presses, such as offset newspaper and packaging print. In *Recollections of Crosfield Electronics*, John Crosfield wrote, "My philosophy was to concentrate our research on new electronic aids for the printing industry, to maximize the use of our electronic 'know-how' on the one hand and our sales contacts in the printing industry on the other. Eventually, we had the greatest range of electronic equipment for the printing industry of any company in the world."

In the 1950s, Crosfield developed a suite of successful automation products for the industry:

- Secatron: an optical system for packaging printers that kept images in the right position on the cardboard so they would look right on the finished carton.
- Webatron: a system like Autotron for high-speed presses that regulated that movement of paper through the press for delivery to folders and sheeters.
- Trakatron: a system for regulating print on web-fed cellophane and wax-paper presses.
- Idotron: a system for measuring ink density on a web press to keep the color reproduction consistent during press runs.
- Viscomex: an ink viscosity control system that added solvents to the ink automatically as needed because of evaporation.
- Flying Paster: an automatic splicing mechanism that enabled presses to go from one roll of paper to the other without slowing down or stopping the press.

Many of these systems relied upon photoelectric cells to detect movement of paper or printed images on the paper. Crosfield's expertise in optical sensors led him to several other important breakthroughs in the composition and preparatory stages of print production. These developments took place in an environment of intense global competition with companies in Europe, the US, and Middle East.

By the 1960s, the printing industry had been moving rapidly into offset lithography. A major factor in this regard was the displacement of hot-metal typesetting with cold type—that is, phototypesetting systems. While Crosfield was not the inventor of the first phototypesetter, his company was a designer and builder of the Lumitype 540 under patents from the original inventors at Photon in the US.

This relationship would continue through the development of the high-speed Photon 713 in 1965, which was the first computer-controlled phototypesetting system.

Among the greatest successes of Crosfield Electronics Ltd. was its color scanning systems. The Crosfield Scanatron—which was developed in 1958—was the first scanning technology that could make color corrections and eliminated the time-consuming work of retouchers.

Crosfield continued with advancements in color scanning throughout the 1960s. The Magnascan was introduced in 1969, and it could scan a color transparency. It also had the software capability to adjust the size, form, color, and hue such that the printed image was of the finest quality anywhere.

While the Magnascan was an international success, it was developed at the same time as Rudolf Hell's Chromograph. Recognizing that a battle over who invented and patented the drum scanner first, the two men signed an agreement giving cross licenses for a modest royalty. Crosfield and Hell remained good friends from that point forward.

In addition to accomplishments in the graphic arts, Crosfield Electronics Ltd. (CEL) also developed computerized business systems and—leveraging the expertise in optical devices—developed a very successful automated banknote sorting and processing technology.

Despite the tremendous success of the company, an attempt to take CEL public in 1974 was made during a collapse of the stock exchange, and Crosfield ended up selling to De La Rue. The scanning business—the most profitable aspect of CEL—was sold by De La Rue in 1989 to a joint venture of Fuji and DuPont called Fujifilm Electronics Imaging.

With the introduction of the desktop PC—and especially the desktop publishing system associated with the Apple Macintosh computer in 1985—the era of Crosfield's domination of graphic arts electronics had ended.

John Crosfield received many accolades for his contributions to the printing industry over nearly five decades, including four UK

Queen's Awards and the gold medal of the Institute of Printing in 1973. He remained a board member of De La Rue until 1985 and thereafter was Honorary President of CEL. A very modest, personable, and generous man, John Crosfield died on March 25, 2012, at his home in Hampstead at age ninety-six.

25

Rudolf Hell

1901–2002

Electronic Engraving and Color Scanning
Kiel, Germany
1950

During the twentieth century, printing technology made a major transition from mechanical to photomechanical and from analog electronic to digital systems. The process took decades, and by the 1990s all the technology of graphic arts production was changed: design, text and image acquisition, typesetting, prepress, printing, binding, and finishing. The result of this innovation was a dramatic improvement in the speed, quality, variety, and complexity of printed material.

It is remarkable how the pace of electronic advancement—beginning with Marconi's December 1901 wireless trans-Atlantic radio broadcast—accelerated throughout the century and influenced every industry and occupation. There were many important theoreticians, scientists, and engineers that participated in this progression. A few of the most familiar names from the early 1900s are Max Planck (quantum theory, 1900), John Fleming (diode, 1904), and Albert Einstein (relativity, 1905).

Rudolf Hell, whose life spanned the entire twentieth century, was an outstanding representative of the generation of engineers who participated in the electronics revolution. Particularly after World

War II, Hell was responsible for many critical inventions related to the image reproduction aspects of printing. The truth is that Rudolf Hell is such an important figure, and his inventions are so numerous—he is credited with more than 130 patents—that it is only possible to focus here on the most significant of his achievements.

Rudolf Hell was born on December 19, 1901, in Eggmühl, Germany, about seventy miles northeast of Munich. Rudolf's father, Karl Hell, was the train stationmaster for the Royal Bavarian Railway at Eggmühl. The family lived in the romantic-style station building, and it was here that Rudolf had early exposure to the telegraph.

Rudolf's mother was the daughter of a brewery owner, and he would later describe her as "vivacious." It seems that Rudolf inherited his entrepreneurialism from his mother since his father was very much "a proper official" and "quite relaxed and Bavarian in character." When Rudolf was six years old, his father relocated the family to the town of Eger—now a border town in the Czech Republic—to become the stationmaster of an Austro-Hungarian freight station, an important transfer point to the Saxon and Bohemian railway lines. Rudolf attended school for twelve years in Eger.

While in high school at the Rudolphinum Oberrealschule, Hell earned high marks in math and science. He would later explain, "I was always the best in physics and in mathematics too. I was mediocre in languages and poor in the subjects that required me to study a lot." Just before age eighteen, Rudolf enrolled in the Technical University at Munich (THM) to study electrical engineering. In 1923, he received a master's degree from THM in electrical engineering at the age of twenty-two. He then became assistant to Professor Max Dieckmann, a specialist in wireless telegraphy at the university. While continuing his studies, Hell worked with Dieckmann on innovations in radio-direction finding and television technology.

Germany is home to the invention of the CRT (cathode ray tube) and the oscilloscope by Karl Ferdinand Braun in 1897. This device is the foundation of the first functioning television systems that were invented simultaneously by Masataka Takayanagi (Japan), Philo Farnsworth (US), and Vladimir Zworykin (USSR and US) in the mid-1920s. For many people, the idea of transmitting moving pictures through a wireless medium was a fantasy. Such was the case with Dieckmann's superiors at THM who forced the professor to rename his course in "wireless television" to something less provocative.

This did not stop Rudolf Hell. In 1925, he filed a patent along with Dieckmann for a photoelectric scanning tube that was basically a primitive television camera. The concept behind this device—the "image-dissector tube"—was that images or scenes could be broken up into small picture elements and transmitted to a receiver for viewing. In that same year, Hell and Dieckmann presented a complete radio-based television system at the Transport Exposition in Munich that included a reception station.

Rudolf Hell's "image dissection," transmission, and reassembly of picture elements (pixels) is at the heart of his remarkable career. To understand the significance of the accomplishment, it is useful to look at it from a modern-day perspective. In 1925, Hell's "image-dissector tube" was understood by a handful of engineers and physicists; today, the concept and its practical application are familiar to billions of people all over the world in the form of megapixel cameras in their mobile phones.

In 1927, Hell received a PhD for his dissertation on "direct-indicating radio-direction finder for aviation." Far ahead of its time, this system enabled pilots to navigate their flights in poor visibility conditions by homing in on a radio beacon. While initially ridiculed by "experts" because "no one flies in the fog anyway and when the weather is clear you don't need it," Hell's invention became the technical basis of all automatic guidance systems and aircraft autopilot technologies. Rudolf Hell received the present-day equivalent of $750,000 from firms in Germany and the US for the license to use his invention in their aviation equipment.

The young Rudolf had always said he did not want to remain in academia as an "ivory tower scientist." He ended his time as assistant to Max Dieckmann, took the money he had earned, and founded his own business called Dr.-Ing. Rudolf Hell Company in May 1929 in Neubabelsberg near Berlin. It was here that Hell began work on a project that would bring him significant worldwide recognition.

On his 1929 patent application, Hell called his invention a "device for electrically transmitting written characters," and it was later renamed the Hell-Schreiber (Hell Recorder). The device—in which originals were broken into dots and electronically transmitted, received, and reassembled—later became the basis of the fax machine. Rudolf sold the patent for his invention to Siemens for the equivalent of $500,000, and he used the money to invest in his business.

At twenty-eight-years-old and married, Hell hired about a dozen employees to work in his company machine shop, design office, laboratory, and business office located in a house that he bought in Berlin-Dahlem. In 1931, Hell expanded the production operations of his company to accommodate the growth and influence of the Hell-Schreiber. By 1934 the device was being used by news agencies across the globe, including the major German agencies, Reuters and TASS.

When World War II started, news organizations and governments alike used the Hell-Schreiber because it was not susceptible to transmission disturbances that were frequent during wartime. By the end of the war, Hell had sold more than fifty thousand units and expanded the offerings of his company to include radio position-finding equipment, radio compasses, and encryption devices.

As a businessman who refused to leave Germany during the war, Rudolf Hell chose to maintain his company and its two factories and one thousand employees throughout the conflict. It is not possible, based on available information, to determine the nature of the relationship of the Dr.-Ing. Rudolf Hell Company with the Nazi regime and German military during the war years. In the end, however, the bombing of Berlin between 1940 and 1945 resulted in the partial destruction of the Hell manufacturing facilities, and the business was lost.

Rudolf Hell declined an offer to relocate his enterprise in Britain after the war and instead elected rebuild in the city of Kiel in the north of Germany. Beginning in 1947, Hell started on a path that would lead to important contributions to electronic graphic-arts technology.

The reestablishment of the Dr.-Ing. Rudolf Hell Company in Kiel was difficult since resources, materials, and tools were in limited supply. Hell's first postwar contract was with Siemens, and he worked on several projects including a fax machine with a spinning drum and a flatbed scanner/printer. These projects were way ahead of their time—Japanese companies, such as Canon, successfully developed and marketed these technologies twenty years later—and Siemens decided to abandon them.

Rudolf Hell's previous discoveries and accomplishments in image dissection led him in the 1950s to explore electronic systems for the graphic arts that would replace the previous generation of mechanical and photomechanical methods.

The following are the most important of Hell's inventions in this field:

- Klischograph (1951): Hell first tested this device that scanned originals and converted them electronically into an engraved printing block. The Klischograph, which was released commercially to the newspaper industry in 1954, dramatically reduced the production time required to make halftone plates by combining three stages of production—film processing, screening, and etching—into one operation.
- Digiset (1956): Different from the technology of other phototypesetting equipment of the era—where complete characters were projected onto photographic material from a negative—the Digiset built each character from digital elements and projected them onto a CRT, and then this image was projected onto photosensitive material. With this system, Rudolf Hell invented the first "bitmap" fonts that would become standard technology in desktop publishing three decades later. The system was launched commercially in 1965.
- Colorgraph (1956): By the late 1950s, technology firms were locked in an international race to invent an electronic scanning system that could produce process color sepa-

rations. Rudolf Hell was well-positioned to compete. He put significant resources into the development of a flatbed scanning system that could convert an original photographic transparency or print into fully color corrected separations all in one step. Hell launched the Colorgraph system commercially in 1958.

- Helio-Klischograph (1961): In October 1959, Hell was approached by a representative of a major German publishing company and asked to devise an electronic system for automated engraving of gravure cylinders. By 1960, Hell had developed a prototype system on a lathe that employed the Klischograph engraving head. Hell's solution—that enabled mass production of gravure cylinders for illustration, decorative, and packaging printing—was debuted at DRUPA in 1962. The Helio-Klischograph system replaced approximately ten separate manual steps in gravure cylinder preparation and is still in use today.

- Chromograph (1963): Throughout the 1960s, Hell perfected the successful scanning and cylinder technologies of the Klischograph and Colorgraph. Striving for the most advanced drum scanning technology of the era, Hell was in a rivalry with John Crosfield of London for the first system to market. By 1967, Rudolf Hell had filed a patent application for his daylight drum-scanning technology, and the Chromograph DC300 system was brought to market in 1970. Hell's drum-scanning technology became the standard for high-quality color reproduction in the printing industry for the next three decades, and well into the first two decades of the twenty-first century.

Altogether these inventions by Rudolf Hell represent a stage in the transition of print technology from the mechanical to the digital age. Later referred to as proprietary systems, these self-contained computerized solutions—and others like Crosfield, Scitex, and Linotype—in the 1950s, '60s, and '70s were the precursors to the desktop publishing revolution of the mid-1980s.

Rudolf Hell, who is sometimes called the Thomas Edison of the graphic arts industry, continued to develop electronic technologies—including the electronic-page composition system called ChromaCom—right up to the fiftieth anniversary of his company in 1979.

Rudolf Hell retired from the management of his company in 1972, but remained active as chairman of the board. In 1981, he sold the business to Siemens and became honorary chairman of the supervisory board. He officially retired in 1989, and the firm was sold again to Linotype, creating the Linotype-Hell Company. The assets of Linotype-Hell were acquired by Heidelberg in 1996, and the technologies pioneered by Rudolf Hell were incorporated into the Heidelberg prepress and press systems.

Hell was the recipient of numerous accolades and honors during his lifetime, including the Gutenberg Award (1977), Werner-von-Siemens Ring in recognition of achievements in the natural sciences and technology (1978), and Honorary Citizen of Kiel (1979), and a Kiel city street was named after him (2001). Rudolf Hell died in Kiel on March 11, 2002, at one hundred years old.

the wonderful
world
of
INSECTS
BY ALBRO GAUL
ILLUSTRATED WITH
PHOTOGRAPHS
the wonderful
world
of
INSECTS
ALBRO GAUL
RINEHART

26

Louis Moyroud
1914–2010

Phototypesetting
Lyon, France
1953

The 1953 edition of *The Wonderful World of Insects* by Albro Tilton Gaul is an important book…but not just because it is about insects. *The Wonderful World of Insects* is also important because it is the first book ever produced with phototypesetting.

Prior to 1953, almost everything printed used the casting method known as "hot type." The mechanized production of molten metal-type characters was first created by Gutenberg in the 1440s, and his technique thrived for over four hundred years. The system was revolutionized by Mergenthaler's Linotype machine in 1880s.

Beginning in the 1940s, with the invention of "cold type" by Louis Moyroud and his fellow inventor Rene Higonnet, the typographic process was again revolutionized. Moyroud and Higonnet's breakthrough was significant because type creation went from being a mechanical to an electronic process. But more fundamentally, the two French engineers had initiated technologies that would later lead to a transformation of the graphic arts from analog to digital technology and this revolution continues to this day.

Louis Marius Moyroud was born on February 16, 1914, in Moirains, Isère, France, and was the only child of Marius and Ann

Marie Vial Moyroud. Louis never knew his father, who died when he was an infant. His mother worked in a textile factory.

As a student, Louis was outstanding. He received government support to study engineering at one of the best institutions in France, École Nationale Supérieure d'Arts et Métiers (Arts et Métiers Paris Tech), and he graduated from there in 1936. Upon graduation, he served in the French army as a second lieutenant and was promoted to first lieutenant in 1939.

Louis's work as an inventor began when a subsidiary of ITT Corporation in Lyon called LMT Laboratories hired him. International Telephone and Telegraph was by this time a global corporation that owned both telephone services infrastructure and manufacturing operations that produced telephone equipment.

In the early 1940s, Louis was working with Rene Alphonse Higonnet when they observed the traditional process of hot-metal typesetting in a French printing plant. Based on what they knew about the science of light, optics, and photography, Moyroud and Higonnet believed that an alternative to the casting of molten-metal typesetting could be developed.

As with previous breakthrough technologies, there were many people trying to displace hot-metal typesetting with a more advanced system. Moyroud and Higonnet were the first to build a functioning solution that was made into a commercial product. Much of the pressure to find a viable photographic typesetting system was being driven by the replacement of the letterpress printing method by offset lithography.

According to Louis's son, Patrick, "My dad always said they thought it was insane [the Linotype process]. They saw the possibility of making the process electronic, replacing the metal with photography. So they started cobbling together typewriters, electronic relays, a photographic disc."

Moyroud and Higonnet worked throughout the war years on the project and first demonstrated their invention in September 1946 in Lyon. Their first functioning photocomposing machine used a typewriter, a strobe light, and a series of lenses to project characters from a spinning disk onto photographic paper. The typeset copy could then be used to make printing plates. Moyroud and Higonnet called their machine the Lumitype.

With the post–World-War-II technology revolution underway, Louis and Rene moved to the United States to pursue the commercialization of their concept. They approached Vannevar Bush, President of Massachusetts Institute of Technology (MIT) and President Roosevelt's top technology adviser, with their prototype. Bush put them in touch with William Garth, President of Lithomat Corporation, a Cambridge, Massachusetts manufacturer of presensitized offset duplicator plates.

Garth was convinced that a successful phototypesetting system would stimulate the growth and expansion of offset printing and drive sales of his Lithomat plates. He formed the Graphic Arts Research Foundation to raise financial resources for the development and marketing of Moyroud and Higonnet's invention. Encouraged by the possibility of dramatic cost reductions in the print-production process, Garth attracted support from major newspaper publishers, book printers, and traditional typesetting services.

After several years of development work, significant support for the project came in. Garth spent over one million dollars to create a prototype phototypesetter. He also changed the name of his firm to Photon, Inc. The prototype device was called Petunia, and it was used to set the type of *The Wonderful World of Insects* in 1953. In 1957, Moyroud and Higonnet were granted a patent for their invention, and tens of thousands of phototypesetting machines were sold.

For the next thirty years, this method of producing type was dominant for printing, publishing, and advertising copy. Mechanical artwork was produced by "paste up" artists around the world for reproduction on offset-lithographic printing presses. Hot-metal type and letterpress printing rapidly receded into the background although some Linotype-generation systems remained in use for specialty work, and that continues today.

In 1985 (two years after the death of Higonnet), Louis Moyroud and Rene Higonnet were inducted into the National Inventors Hall of Fame in Alexandria, Virginia. Ironically, 1985 is also the year of the advent of desktop publishing, a technology that would, within a few years, completely displace phototypesetting as a method of producing type for print. This fact shows that phototypesetting was a significant evolutionary development along the path of the digital transformation of the graphic arts.

Louis Moyroud continued his work on phototypesetting systems into the 1980s, and his career as an inventor extended beyond the displacement of his most important contribution. He retired to Delray Beach, Florida, where he later died in June 2010 at the age of ninety-six.

Frank Romano, who worked with Louis Moyroud and was the advertising manager of Photon in 1969, wrote the following tribute to Louis, "He had a wonderful sense of humor and an unassuming demeanor. He had collected most of the early phototypesetters and donated them to the Museum of Printing in North Andover. Petunia is on display. John Crosfield, Rudolf Hell, Benny Landa, and Dan Gelbart are among the inventors who moved the printing industry to new levels, but the era of automation began with Louis and Rene. Louis is now gone, and the revolution he began is now ended. But other revolutions continue."

Helvetica

Aa Ee Rr
Aa Ee Rr

a

Buchdruck

a b c d e f g h i j k l m
n o p q r s t u v w x y z
0 1 2 3 4 5 6 7 8 9

27

—●—

Eduard Hoffmann

1892–1980

Helvetica Typeface
Münchenstein, Switzerland
1957

It is a fact of typographic history that the font Helvetica exists today because of Hamburgers...not the food, but the word "Hamburgers." Eduard Hoffmann, the creator of the ubiquitous typeface, knew the word "Hamburgers" contained the complete range of character attributes in the alphabet; he knew that from this one word, the quality of a typeface design could be evaluated, that the features of its anatomy could be examined.

And so early in the design of the precursor to Helvetica—called Neue Haas Grotesk—Eduard Hoffmann of the Swiss-based Haas Type Foundry wrote to his designer and confidante, Max Miedinger, "But our first priority is the word 'Hamburgers.' It is the universal type founders' word that contains all the varieties of letters."

Helvetica is probably the most successful typeface in all of history. It is everywhere all the time, and there are reasons for this: Helvetica is neutral and easy to read; its different weights and styles effectively embody almost any meaning or message. Helvetica is plain but it is also modern.

Helvetica came about when typography and printing technology were moving from the metal-casting, mechanical, and letterpress

221

era to the electronic typesetting, word processing, laser imaging, and computer age. It rode atop this transformation and became the first truly international typeface. By the mid-1960s, Helvetica emerged as a global standard for public signage, corporate identity, and communications.

Regardless of one's opinion of the aesthetics and usefulness of Helvetica today, its creation and development—the people who designed it and how they developed it—is one of the most interesting and important accomplishments of twentieth-century graphic arts.

Eduard Hoffmann was born on May 26, 1892, in Zurich, Switzerland. At school, he studied technology and engineering in Zurich, Berlin, and Munich with a specific interest in aviation. In 1917, the twenty-five-year-old Hoffmann took a position under the direction of his uncle, Max Krayer, at Haas'sche Schriftgiesserei (Haas Type Foundry) in Münchenstein, Switzerland, and made a commitment to the profession of typography. In 1937, Eduard became comanager of the company with Krayer and, after his uncle's death in 1944, became sole manager where he remained until his retirement in 1965.

As early as 1950, Hoffmann planned to introduce a new sans serif typeface into the Swiss market that could compete with those coming from other European countries. The origins of sans serif typefaces date back to the late eighteenth century and where it was used with an embossing technique to enable the blind to read. The first fully developed sans serif (also known as grotesque or grotesk) made its appearance in Germany around 1825, and a French type founder first used the term sans serif (without decorative extensions) in 1830.

The sans serif types that Hoffmann wanted to compete against became popular and successful in the late nineteenth century. This was true of Akzidenz Grotesk of the Berlin-based H. Berthold AG type foundry, for example, which was originally designed in 1896. But there were neo-grotesk faces that had since entered the market, Bauer's Folio and Frutiger's Univers for example, that threatened to eclipse Hoffmann's venture.

In 1956, as grotesk fonts were surging in Europe, Hoffmann thought the timing was right to attempt a specifically Swiss variety. He contacted Max Miedinger, who had been a salesman and type designer at the Haas Type Foundry for the previous ten years, and wrote, "He was the only man to design a new typeface for Haas."

Miedinger's role in the process was decisive, and many credit him more than Hoffmann for the creation of Helvetica. It is true that Miedinger made the original hand-drawn letters of the alphabet. But the aesthetic component was only one side of the value that Miedinger brought to Haas. It was his in-depth knowledge and relationship with the customers of the type foundry that made Miedinger indispensable to the success of Neue Haas Grotesk.

Miedinger had access to some of the most brilliant Swiss graphic artists as well as advertising representatives from major Swiss corporations—the chemical firm J. R. Geigy AG among them—and through a painstaking and collaborative development process headed up by Hoffmann, Neue Haas Grotesk took shape.

Throughout 1957 and 1958, the two men collaborated back and forth, fine-tuning each character. The record of the exchange between Hoffmann and Miedinger has been preserved and can be followed in detail in the book *Story of a Typeface: Helvetica Forever.* The book includes photographic reproductions of the letters the men wrote to each other as well as Hoffmann's project notebook.

Hoffmann knew that designing a great typeface was not only about the beauty and logical construction of each individual character even though this was an important aspect. Each character had to fit together with all the other characters in the various combinations that make letters into words. There was also the technical question of how the typeface would look at different sizes and once it was printed with ink on paper.

As Hoffmann explained in 1957, "Praxis has shown that a new typeface cannot be correctly and objectively evaluated until it is in printed form. But even then, it is quite curious to find that a letter might be very satisfactory in a word while seemingly quite out of place in another context. This makes it necessary to consider its design anew, which usually leads to unavoidable compromises."

Once they were satisfied with the basic letterforms and had designed enough weights and sizes—at that time, the Haas Type Foundry was punch cutting, engraving, and typecasting by hand thousands upon thousands of individual characters in metal—the men took their product to market. With the help of some well-designed promotional brochures and an initial launch at the Graphic 57 trade fair in Lausanne, Neue Haas Grotesk became a hit. By 1959, about 10 percent of the printers in Switzerland were carrying it.

The Haas Type Foundry was majority-owned by the German firm D. Stempel AG. In turn, Stempel was in a contract with the multinational Linotype Corporation to produce machine-manufactured metal typeforms. To expand the appeal of Hoffmann and Miedinger's typeface and to bring it to the world of mass-production typography, especially in the US, Linotype's marketing department pushed for Neue Haas Grotesk to be renamed.

Linotype initially suggested that it simply be named Helvetia (Latin for Switzerland). Hoffmann felt that although it was distinctly a Swiss product, the typeface could not have the exact same name as the country. He came up with Helvetica, which means "the Swiss typeface," and all involved accepted the new name developed by its creator.

Into the 1960s, Helvetica gained spectacular popularity and was adopted as the in-house typeface of various international corporations, many of which still use it to this day. Commentary on the significance and social driving force behind the success of Helvetica has often referred to postwar economic expansion. There was a thirst in the 1950s within the creative community for visual clues that conveyed optimism about the future. Designers wanted an excessively modern look that helped to put the bad memories of the world wars of the first half of the twentieth century far behind. For many, Helvetica accomplished this goal.

In 1971, Hoffmann established a foundation with the aim of creating a museum dedicated to the printing industry. In 1980, in a former Gallician paper mill on the Rhine, a museum opened with Hoffmann's collection of papers on the history of the Haas Type Foundry as one of its main attractions. Eduard Hoffmann died in Basel, Switzerland, on September 17, 1980.

OCR B

Aa Ee Gg
Nn Qq Rr a

Electronic

abcdefghijklm
nopqrstuvwxyz
0123456789

28

Adrian Frutiger
1928–2015

Univers and OCR-B Typefaces
Paris, France
1957

Adrian Frutiger was one of the most important type designers of his generation, having created some forty fonts, many of them still widely used today. He was also a teacher, author, and specialist in the language of graphic expression and—since his career spanned metal, photomechanical, and electronic-type technologies—Frutiger became an important figure in the transition from the analog to the digital eras of print communications.

Frutiger was born on May 24, 1928, in the town of Interseen, near Interlaken and about sixty kilometers southeast of the city of Bern, Switzerland. His father was a weaver. As a youth, Adrian showed an interest in handwriting and lettering. He was encouraged by his family and secondary schoolteachers to pursue an apprenticeship rather than a fine-arts career.

At age sixteen, Adrian obtained a four-year apprenticeship as a metal-type compositor with the printer Otto Schlaeffli in Interlaken. He also took classes in drawing and woodcuts at a business school in the vicinity of Bern. In 1949, Frutiger transferred to the School of Applied Arts in Zurich, where he concentrated on calligraphy. In 1951, he created a brochure for his dissertation entitled "The

Development of the Latin Alphabet" that was illustrated with his own woodcuts.

It was during his years in Zurich that Adrian worked on sketches for what would later become the typeface Univers, one of the most important contributions to postwar type design. In 1952, following his graduation, Frutiger moved to Paris and joined the foundry Deberny & Peignot as a type designer.

During his early work with the French type house, Frutiger was engaged in the conversion of existing metal-type designs for the newly emerging phototypesetting technologies. He also designed several new typefaces—Président, Méridien, and Ondine—in the early 1950s.

Sans serif type is a product of the twentieth century. Also known as grotesque (or grotesk), sans serif fonts emerged with commercial advertising, especially signage. The original sans serif designs (beginning in 1898) had qualities—lack of lowercase letters, lack of italics, the inclusion of condensed or extended widths, and equivalent cap and ascender heights—that seemingly violated the rules of typographic tradition. As such, these early sans serif designs were often considered too clumsy and inelegant for professional type houses and their clients.

Along with the modern art-and-design movements of the early twentieth century, a reconsideration of the largely experimental work of the first generation of sans serif types began in the 1920s. Fonts, such as Futura, Kabel, and Gill Sans, incorporated some of the theoretical concepts of the Bauhaus and DeStijl movements and pushed sans serif to new spheres of respectability.

However, these fonts—which are still used today—failed to elevate sans serif beyond headline usage and banner advertising and into broader application. Sans serif type remained something of an oddity and not yet accepted by the traditional foundry industry as viable in terms of either the style or legibility of serif fonts.

In the 1930s, especially within the European countries that fell to dictatorship prior to and during World War II, there was a backlash against modernist conceptions. Sans serif type came under attack, was derided as "degenerate," and banned in some instances.

Exceptions to this trend were in the US, where the use of grotesque types was increasing, and Switzerland, where the minimalist typographic ideas of the Bauhaus were brought by designers who had fled the countries ruled by the Nazis.

After the war, interest in sans serif type design was renewed as a symbol of modernism and a break from the first four decades of the century. By the late 1950s, the most successful period of sans serif type opened up, and the epicenter of this change emerged in Switzerland, signified by the creation of Helvetica (1957) by Eduard Hoffmann and Max Miedinger of the Haas Type Foundry in Münchenstein.

It was the nexus of the creative drive to design the definitively "modern" typeface and the possibilities created by the displacement of metal type with phototypesetting that brought sans serif from a niche font into global preeminence.

This was the cultural environment that influenced Adrian Frutiger as he set about his work on a new typeface as a Swiss-trained type designer at a French foundry. As Frutiger explained in a 1999 interview with *Eye Magazine*, "When I came to Deberny & Peignot in Paris, Futura (though it was called Europe there) was the most important font in lead typesetting. Then one day, the question was raised of a grotesque for the Lumitype-Photon [the first phototypesetting system]....

"I asked him [Peignot] if I might offer an alternative. And within ten days, I constructed an entire font system. When I was with Käch, I had already designed a thin, normal, semi-bold and italic Grotesque with modulated stroke weights. This was the precursor of Univers.... When Peignot saw it, he almost jumped in the air. 'Good heavens, Adrian, that's the future!'"

Originally calling his type design Monde (French for world), Frutiger's innovation was that he designed twenty-one variations of Univers from the beginning; for the first time in the history of typography, a complete set of typefaces were planned precisely as a coherent system. He also gave the styles and weights a numbering scheme beginning with Univers 55. The different weights (extended, condensed, ultra-condensed, etc.) were numbered in increments of ten—such as, 45, 55, 65, etc.—and styles with the same line thickness were numbered in single digit increments such as 53, 54, 55, 56, etc.

Univers was released by Deberny & Peignot in 1957, and it was quickly embraced internationally for both text- and display-type purposes. Throughout the 1960s and '70s, like Helvetica, it was widely used for corporate identity (GE, Lufthansa, Deutsche Bank). It was the official promotional font of the 1972 Munich Olympic Games.

Frutiger explained the significance of his creation in the interview with *Eye Magazine*, "It happened to be the time when the big advertising agencies were being set up, they set their heart on having this diverse system. This is how the big bang occurred and Univers conquered the world. But I don't want to claim the glory. It was simply the time, the surroundings, the country, the invention, the postwar period, and my studies during the war. Everything led towards it. It could not have happened any other way."

Had Adrian Frutiger retired at the age of twenty-nine after designing Univers, he would have already made an indelible contribution to the evolution of typography. However, his work was by no means complete. By 1962, Frutiger had established his own graphic design studio with Bruno Pfaffli and Andre Gurtler in Arcueil near Paris. This firm designed posters, catalogs, and identity systems for major museums and corporations in France.

Throughout the 1960s, Frutiger continued to design new typefaces for the phototypesetting industry, such as Lumitype, Monotype, Linotype, and Stempel AG. Among his most well-known later sans serif designs were Frutiger, Serifa, and Avenir. Frutiger's font systems can be seen to this day on the signage at Orly and Charles de Gaulle airports and the Paris Metro.

The penetration of computers and information systems into the printing and publishing process were well underway by the 1960s. In 1961, thirteen computer and typewriter manufacturers founded the European Computer Manufacturers Association (ECMA) based in Geneva. A top priority of the EMCA was to create an international standard for optical character recognition (OCR)—a system for capturing the image of printed information and numbers and converting them into electronic data—especially for the banking industry.

By 1968, OCR-A was developed in the US by American Type Founders—a trust of twenty-three American type foundries—and it was later adopted by the American National Standards Institute (ANSI). This was the first practically adopted standard mono-spaced font that could be read by both optical machines and the human eye.

However, in Europe, the ECMA wanted a font that could be used as an international standard such that it accommodated the requirements of the typographic needs and computerized scanning technologies all over the world. Among the issues, for example, were the treatment of the British pound symbol (£) and the Dutch IJ and French oe (œ) ligatures. Other technical considerations included the ability to integrate OCR standards with typewriter and letterpress fonts in addition to the latest phototypesetting systems.

In 1963, Adrian Frutiger was approached by representatives of the ECMA and asked to design OCR-B as an international standard with a non-stylized alphabet that was also aesthetically pleasing to the human eye. Over the next five years, Frutiger exhibited the exceptional ability to learn complicated technical requirements of engineers, such as the grid systems of the different optical readers, the strict spacing requirements between characters, and the special shapes needed to make one letter or number distinguishable from another.

In 1973, after multiple revisions and extensive testing, Adrian Frutiger's OCR-B was adopted as an international standard. Today, the font can be found on UPC barcodes, ISBN barcodes, government-issued ID cards, and passports. Frutiger's OCR-B font will no doubt live on into the distant future—alongside various 2D barcode systems—as one of the primary means of translating analog information into electronic data and back again.

Adrian Frutiger's type design career extended well into the era of desktop publishing, PostScript fonts, and the Internet age. In 1989, Frutiger published the English translation of *Signs and Symbols: Their Design and Meaning*, a theoretical and retrospective study of the two-dimensional expression of graphic drawing with typography among its most advanced forms. For someone who spent his life working on the nearly imperceptible detail of type and graphic design, Frutiger exhibited an exceptional grasp of the historical and social sources of man's urge toward pictographic representation and communication.

As an example, Frutiger wrote in the introduction to his book, "For twentieth century humans, it is difficult to imagine a void, a chaos, because they have learned that a kind of order appears to prevail in both the infinitely small and the infinitely large. The understanding that there is no element of chance around or in us, but that all things both mind and matter, follow an ordered pattern, supports the argument that even the simplest blot or scribble cannot exist by pure chance or without significance but rather that the viewer does not clearly recognize the causes, origins, and occasion of such a 'drawing.'"

Adrian Frutiger died on September 10, 2015, at the age of eighty-seven.

elevate the soul
reter of wis
ALL ARTS an
er is a

29

Hermann Zapf

1918–2015

Digital Typography
Darmstadt, Germany
1964

O n June 12, 2014, Apple released its San Francisco system font for OSX, iOS, and watchOS. Largely overlooked amid the media coverage of other Apple product announcements, the introduction of the San Francisco font was a noteworthy technical event.

San Francisco is a neo-grotesk, sans serif, and Pan European typeface with characters in Latin as well as Cyrillic and Greek scripts. It is significant because it is the first font to be designed specifically for all of Apple's display technologies. Important variations have been introduced into San Francisco to optimize its readability on Apple desktop, notebook, TV, mobile, and watch devices.

It is also the first font designed by Apple in two decades. San Francisco extends Apple's association with typographic innovation that began in the mid-1980s with desktop publishing. From a broader historical perspective, Apple's new font confirms of the ideas developed more than fifty years prior by renowned calligrapher and type designer Hermann Zapf. Sadly, Zapf died at the age of ninety-six on June 4, 2015, just one week before Apple's San Francisco announcement.

Hermann Zapf's contributions to typography are extensive and astonishing. He designed more than two hundred typefaces—the popular Palatino (1948), Optima (1952), Zapf Dingbats (1978), and Zapf Chancery (1979) among them—including fonts in Arabic, Pan-Nigerian, Sequoia, and Cherokee. Meanwhile, Zapf's exceptional calligraphic skills were such that he famously penned the Preamble of the Charter of the United Nations in four languages for the New York Pierpont Morgan Library in 1960.

While he made many extraordinary creative accomplishments—far too many to list here—Hermann Zapf's greatest legacy is the way he thought about type and its relationship to technology. Herman Zapf was among the first and perhaps the most important typographer to theorize about the need for new forms of type driven by computer and digital technologies.

Hermann Zapf was born in Nuremburg on November 8, 1918, during the turbulent times near the end of World War I. As he wrote later in life, "On the day I was born, a workers' and soldiers' council took political control of the city. Munich and Berlin were rocked by revolution. The war ended, and the Republic was declared in Berlin on 9 November 1918. The next day, Kaiser Wilhelm fled to Holland." One wonders if printing employees in Munich and Berlin engaged in actions similar to those of the typographers and lithographers in Russia in 1905. In that great democratic movement, which ultimately resulted in the Russian Revolution of 1917, the typographers at Sytin's printworks in Moscow sparked the creation of community councils that spread across the country and directly challenged the political authority of the absolutist monarchy.

At school, Hermann took an interest in technical subjects. He spent time in the library reading scientific journals and at home, along with his older brother, experimenting with electronics. He also tried hand lettering and created his own alphabets. Hermann left school in 1933 with the intention of becoming an engineer. However, economic crisis and new upheavals in Germany—including the temporary political detention of his father in March 1933 at the prison camp in Dachau—prevented him from pursuing his plans.

Barred from attending the Ohm Technical Institute in Nuremberg for political reasons, Hermann sought an apprenticeship in lithography. He was hired in February 1934 to a four-year apprenticeship as a photo retoucher by Karl Ulrich and Company. In 1935, after reading books by Rudolf Koch and Edward Johnson on lettering and illuminating techniques, Zapf taught himself calligraphy. When management saw the quality of his lettering, the Ulrich firm began to assign him work outside of his retouching apprenticeship.

Zapf refused to take the journeyman's test on the grounds that his training had been interrupted by many unrelated tasks. He never received his journeyman's certificate and left Nuremburg for Frankfurt to find work.

Zapf started his career in type design at the age of twenty after being employed at the Fürsteneck Workshop House, a printing establishment run by Paul Koch, the son of Rudolf Koch. As he later explained, "It was through the print historian Gustav Mori that I first came into contact with the D. Stempel AG type foundry and Linotype GmbH in Frankfurt. It was for them that I designed my first printed type in 1938, a fraktur type called 'Gilgengart.'" Fraktur typefaces are decorative emulations of the calligraphic styles originally developed in Germany in the sixteenth century. A common modern-day use of Fraktur fonts is in the nameplates of many newspapers such as the *New York Times*.

Hermann Zapf was conscripted in 1939 and called up to serve in the German army near the town of Pirmasens on the French border. After a few weeks, he developed heart trouble and was transferred from the hard labor of shovel work to the writing room where he composed camp reports and certificates. When World War II started, Hermann was dismissed for health reasons. In April 1942, he was called up again, this time for the artillery. Hermann was quickly reassigned to the cartographic unit where he became well-known for his exceptional map-drawing skills. He was the youngest cartographer in the German army through the end of the war.

Zapf was captured by the French and held in a field hospital in Tübingen. As he recounted, "I was treated very well, and they even let me keep my drawing instruments. They had a great deal of respect for me as an 'artiste'… Since I was in very poor health, the French sent me home just four weeks after the end of the war. I first went back to my parents in my hometown of Nuremberg, which had suffered terrible damage." In the years following the war, Hermann taught and gave lessons in calligraphy in Nuremberg. In 1947, he returned to Frankfurt and took a position with the Stempel AG foundry with little qualification other than his sketchbooks from the war years.

From 1948 to 1950, while he worked at Stempel on typography designs for metal punch cutting, he developed a specialization in book design. Hermann also continued to teach calligraphy twice a week at the Arts and Crafts School in Offenbach. It was during these years that Zapf designed Palatino and Optima. Working closely with the punch cutter August Rosenberg, Hermann design Palatino and named it after the sixteenth century Italian master of calligraphy, Giambattista Palatino. In the Palatino face, Zapf attempted to emulate the forms of the great humanist typographers of the Renaissance.

Optima, on the other hand, expressed more directly the genius of Zapf's vision and foreshadowed his later contributions. Optima can be described as a hybrid serif-and-sans-serif typeface because it blends features of both: serif-less thick and thin strokes with subtle swelling at the terminals that suggest serifs. Zapf designed Optima during a visit to Italy in 1950 when he examined inscriptions at the Basilica di Santa Croce in Florence. It is remarkably modern yet clearly derived from the Roman monumental capital model.

By the time Optima was released commercially by Stempel AG in 1958, the industry had begun to move away from metal-casting methods and into phototypesetting. As many of his most successful fonts were reworked for the new photomechanical processes, Zapf recognized—perhaps before and more profoundly than most—that the technology of typography was on an evolutionary path from analog to fully digital systems.

To grasp the significance of Zapf's work, it is important to understand that although *cold* photo type was an advance over *hot* metal type, both are analog technologies—that is, they require the transfer of "master" shapes from manually engraved punches or hand-drawn outlines to final production type by way of molds or photomechanical processes.

Due to the inherent limitations of metal and mechanical media, analog type masters often contain design compromises. Additionally, the reproduction from one master generation to the next has variations and inconsistencies connected with the craftsmanship of the individual creating the punch cutting or outline drawing.

With digital type, the character shapes exist as electronic files that *describe* fonts in mathematical vector outlines or in raster images plotted on an XY coordinate grid. With computer font data, typefaces have many nuances and features that could never be rendered in metal or photo type. Meanwhile, digital font masters can be copied precisely without any quality degradation from one generation to the next.

From the earliest days of computers, Hermann Zapf began advocating for the advancement of digital typography. He argued that type designers needed to take advantage of the new technologies and needed to create types that reflected the age. Zapf also combined knowledge of the rules of good type design with a recognition that fonts needed to be created specifically for electronic displays which, at that time, existed as CRT-based monitors and televisions.

In 1959, at the age of forty-one, Zapf wrote in an industry journal, "It is necessary to combine the purpose, the simplicity, and the beauty of the types, created as an expression of contemporary industrial society, into one harmonious whole. We should not seek this expression in imitations of the Middle Ages or in revivals of nineteenth-century material as sometimes seems the trend. The question for us is satisfying tomorrow's requirements and creating types that are a real expression of our time but also represent a logical continuation of the typographic tradition of the western world."

Despite a cold reception in Germany—his ideas about computerized type were rejected as unrealistic by the Technical University in Darmstadt where he was a lecturer and by leading printing industry representatives—Hermann persevered. Beginning in the early 1960s, Zapf delivered a series of lectures in the US that were met with enthusiasm.

For example, a talk he delivered at Harvard University in October 1964 became so popular that it led to an offer for a professorship at the University of Texas in Austin. The governor even also made Hermann an "Honorary Citizen of the State of Texas." In the end, Zapf turned down the opportunity due to family obligations in Germany.

Among his many digital accomplishments are the following:

- When digital typography was born in 1964 with the Digiset system of Rudolf Hell, Hermann Zapf was involved. By the early 1970s, Zapf created some of the first fonts designed specifically for any digital system: Marconi, Edison, and Aurelia.

- In 1976, Hermann was asked to head a professorship in typographic computer programming at Rochester Institute of Technology (RIT) in Rochester, New York, the first of its kind in the world. Zapf taught at RIT for ten years and was able to develop his conceptions in collaboration with computer scientists and representatives of IBM and Xerox.

- In 1977, Zapf partnered with graphic designers Herb Lubalin and Aaron Burns and founded Design Processing International, Inc. (DPI) in New York City. The firm developed software with menu-driven typesetting features that could be used by nonprofessionals. The DPI software was focused on automating hyphenation and justification as opposed to the style of type design.

- In 1979, Hermann began a collaboration with Professor Donald Knuth of Stanford University to develop a font that was adaptable for mathematical formulae and symbols.

- In the 1990s, Hermann Zapf continued to focus on the development of professional typesetting algorithms with his "hz-program" in collaboration with Peter Karow of the font company URW. Eventually, the Zapf composition engine was incorporated by Adobe Systems into the InDesign desktop publishing software.

Hermann Zapf actively participated—into his '70s and '80s—in some of the most important developments in type technology in the post-war era. This was no accident. He possessed both a deep knowledge of the techniques and forms of type history and a unique appreciation for the impact of information technologies on the creation and consumption of the written word.

In 1971, Zapf gave a lecture in Stockholm called "The Electronic Screen and the Book" where he said, "The problem of legibility is as old as the alphabet for the identification of a letterform is the basis of its practical use.... To produce a clear, readable text that is pleasing to the eye and well-arranged has been the primary goal of typography in all the past centuries. With a text made visible on a CRT screen, new factors for legibility are created." More than forty years before the Apple design team set out to create a font that is legible on multiple computer screens, the typography visionary Hermann Zapf was theorizing about the very same questions.

30

Marshall McLuhan
1911–1980

Multimedia Philosophy
Toronto, Ontario
1964

Marshall McLuhan, the Canadian educator and communications scholar, was born on July 21, 1911. McLuhan became a celebrity in the 1960s for his controversial media studies and peculiar perspective on television and its societal impact.

While readers are likely familiar with McLuhan's aphorism "the medium is the message," they may not know that he was an expert on printing. Much of his writing deals with print-media, its history, and significance as a cultural form, especially the book. One of his most important titles, *The Gutenberg Galaxy: The Making of Typographic Man* (1962), examines prints' contribution to the transformation of mankind's self-image and consciousness during the Renaissance and beyond.

McLuhan was among the first to foresee the coming of electronic media. He had a prophetic view of the information age, one that anticipated the World Wide Web and digital publishing. His relevance to modern media studies was shown by the conferences held worldwide on the centenary of his birth.

Herbert Marshall McLuhan was born in Edmonton, Alberta. His family moved to Winnipeg, Manitoba, after World War I. In

1928, Marshall entered the University of Manitoba where earned a degree in Master of Arts in English. He went on to the University of Cambridge where he earned a PhD in 1943. Marshall married Corrine Lewis in 1939, and they had six children.

McLuhan and his family moved to Toronto in 1946 where he joined the faculty of St. Michael's College of the University of Toronto. In the 1950s, he began the Communication and Culture seminars, gaining a reputation as a media expert, and in 1963, the university created with him the Centre for Culture and Technology.

McLuhan became internationally known with the publication of *Understanding Media: The Extensions of Man* (1964). He advanced the profound idea that "the 'message' of any medium or technology is the change of scale or pace or pattern that it introduces into human affairs." He examined many forms of media: the written word, the printed word, the photograph, the telegraph, the typewriter, the telephone, the phonograph, movies, radio, and television and showed how the content of one media is always another media form.

Marshall McLuhan suffered a stroke in 1979 that affected his speech. The University of Toronto attempted to close his research center shortly thereafter but was prevented by protests, most notably by film director Woody Allen. McLuhan never recovered and died on December 31, 1980.

Since McLuhan's theoretical concepts are difficult to explain, he does it best himself. Below are quotations taken from several TV interviews:

- On "hot" and "cool" media, 1964

    "It has to do with the slang phrase 'the hot and the cool'… 'Cool' in the slang form has come to mean involved, deeply participative, deeply engaged, everything that we had formerly meant by heated… Though the idea that 'cool' has reversed its meaning I think has some bearing on the fact that our culture has shifted its stress on the demand that we become more committed, more involved."

- On the future of publishing, 1966

    "Instead of going out and buying a book of which there have been five thousand copies printed, you will go to the telephone and describe your interests, your needs, your problems... They will at once Xerox, with the help of computers from the libraries of the world, all the latest material just for you, personally, not as something to be put out on the bookshelf. They send you the package as a direct personal service."

- On instantaneous, simultaneous information, 1976

    "At the speed of light, there is no sequence, everything happens at one instant... We live in a world where everything is supposed to be lineal, one thing at a time, connected and logical, goal-oriented. We are now living in a world which pushes the right hemisphere (of the brain) way up...is making the old left hemisphere world, which is our educational establishment, our political establishment, look very foolish."

Admittedly, McLuhan's academic style and references to historical and cultural artifacts make him difficult to read. However, he has made a unique contribution to an understanding of media and how it impacts the cognitive functions and social organization of man. We should embrace and study Marshall McLuhan as the first philosopher of our multimedia world.

ITC Avant Garde Gothic

Aa Ee Rr
Aa Ee Rr

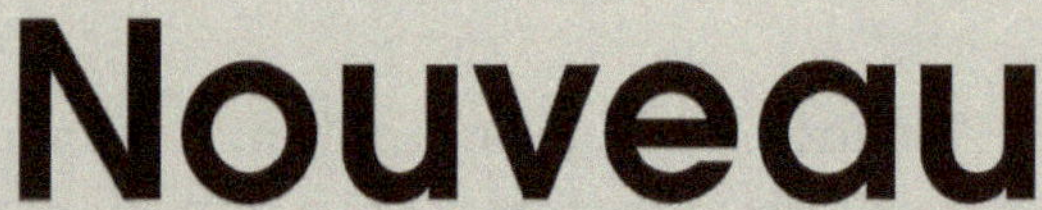

Nouveau

abcdefghijklm
nopqrstuvwxyz
0123456789

31

Herb Lubalin

1918–1981

Avant-Garde Gothic Typeface
New York, New York
1970

Herbert F. Lubalin was born in New York City on March 17, 1918. As a high school student, he did not show a particular interest in the graphic arts although he liked to draw. He entered art school at Cooper Union at the age of seventeen where his interest in typography was nurtured. Lubalin graduated in 1939 and first worked as a freelance designer and typographer. It has been reported that he was fired from a position at a display company after he requested a two-dollar raise on his weekly eight-dollar salary.

Soon thereafter, and for the next twenty-five years, Lubalin worked as an art director for advertising agencies. The New York City firms he worked for included Deutsch & Shea, Fairchild Publications, Reiss Advertising, and Sudler & Hennessey. During these years, Lubalin established himself as a genius of what would be later called "typographics" or "expressive typography,"—that is, words and letters as imagery that express verbal and conceptual meaning.

This was achieved through a meticulous creative approach to advertisements, trademarks and logos, posters, magazines, and packaging design. In 1952, Lubalin won a New York Art Directors Club

Gold Medal as creative director at Sudler & Hennessey, the first of hundreds of awards he would receive throughout his career.

After leaving Sudler in 1964, he established his own graphic design consultancy called Herb Lubalin, Inc. This was the first of multiple businesses and subsidiaries that Lubalin founded in both the US and Europe over the next two decades. In 1970, along with Aaron Burns and Edward Ronthaler, Lubalin created the International Typeface Corporation (ITC), one of the world's first type foundries that had no history in hot-metal type design.

Herb Lubalin achieved worldwide success as an art director and graphic designer during the "Mad Men" era (of the popular AMC TV series) of advertising. Lubalin became identified with graphic clarity and simplicity embodied in the following statement he made some years later, "Typography is a servant—the servant of thought and language to which it gives visible existence."

In terms of the technology of type, this was the age of photo-typesetting. The replacement of hot type with cold type meant that a new library of modern fonts could be developed. It also meant that typeforms could be manipulated in ways that were extremely diffi-cult, if not impossible, in the metal casting process.

Although Lubalin's ITC took up the task of preserving and reviving old classic faces, such as ITC Bookman and ITC Garamond, the foundry also specialized in modern sans serif fonts, such as ITC Franklin, ITC American Typewriter, ITC Kabel, and ITC Bauhaus among many others.

Herb Lubalin's relationship with Ralph Ginzburg—who was convicted in 1963 for violating US obscenity laws—is noteworthy. The two worked together on three groundbreaking magazines: *Eros* (1962), *Fact* (January 1964–August 1967), and *Avant-Garde* magazine.

Avant-Garde became extremely popular in certain creative circles, including among New York's advertising and editorial art directors. Most importantly, *Avant-Garde* was a breakthrough publication. During its four years of existence, the magazine was considered the cutting edge of graphic design, especially its treatment of typography.

Avant-Garde magazine proved to be most significant for Lubalin, specifically for his design of the publication nameplate. The *Avant-Garde* moniker became so popular that Lubalin, his partner, Tom Carnase, and the type designer, Edward Benguiat, developed an entire font set from it. What became the Avant Garde Gothic type design included a series of ligatures that is, combinations of two letters into one type element, an innovative development for a sans serif font.

Officially launched by ITC in 1970, Avant Garde Gothic became one of the most popular typefaces of the era. Although it came under criticism and was eschewed by the post-modernist graphic-design community for its structural and grid-like consistency, Avant Garde Gothic was eventually included in the set of thirty-five base fonts on the Adobe PostScript print engine that was launched in the 1980s. For this reason, Avant Garde Gothic continues to be one of the most popular and often used alternatives to Helvetica.

Herb Lubalin designed some of the most memorable and lasting images of expressive typography that have ever been created. His publication nameplate for "Mother & Child," logo for L'eggs, and logo for the World Trade Center are part of iconic graphic-design history. Lubalin had a near legendary reluctance to talk with anyone, especially the media and trade publications, about his work, and some interpreted his reserved character as a lack of intellectual acumen. However, Lubalin had a very sharp theoretical approach to his craft, and he was not averse to sharing his knowledge with those who wanted to learn, particularly students.

In 1973, Lubalin launched and became editor and art director of International Typeface Corporation's quarterly in-house publication called *U&lc* (Upper and lower case). The journal was an instant force in the industry and rapidly built up a subscription circulation of 170,000 readers. It was in *U&lc* that some of Lubalin's conceptions about graphic and type design can be studied and learned about today.

The following statement—published in the introduction to *Graphis Annual 65/66*—shows that Herb Lubalin possessed a critical and iconoclastic attitude to the industry that he devoted his life to:

> Advertising in the USA is a fairly stupid business. We have made it that way by underestimating the intelligence of the American people. The bulk of our output is devised to appeal to the sub-teenage mentality of that great, big, consuming monster that we have created. Who's responsible? Those of us that put absolute faith in antiquated, ineffective, stereotyped, outmoded, unreliable, unbelievable, valueless research methods, such as copy testing.... If recent statistics are any indication of the value of copy testing, we would all be advised to spend our research money researching successful art directors and copywriters, knowledgeable creative people who have made their reputations not by fancy words and pretty designs but by creating intelligent advertising that appeals to a surprisingly intelligent audience (the American people).

Beginning in 1972, Lubalin began teaching graphic design at Cornell University, and starting in 1976, he taught a course at Cooper Union where he remained until his death on May 24, 1981, at New York University Hospital.

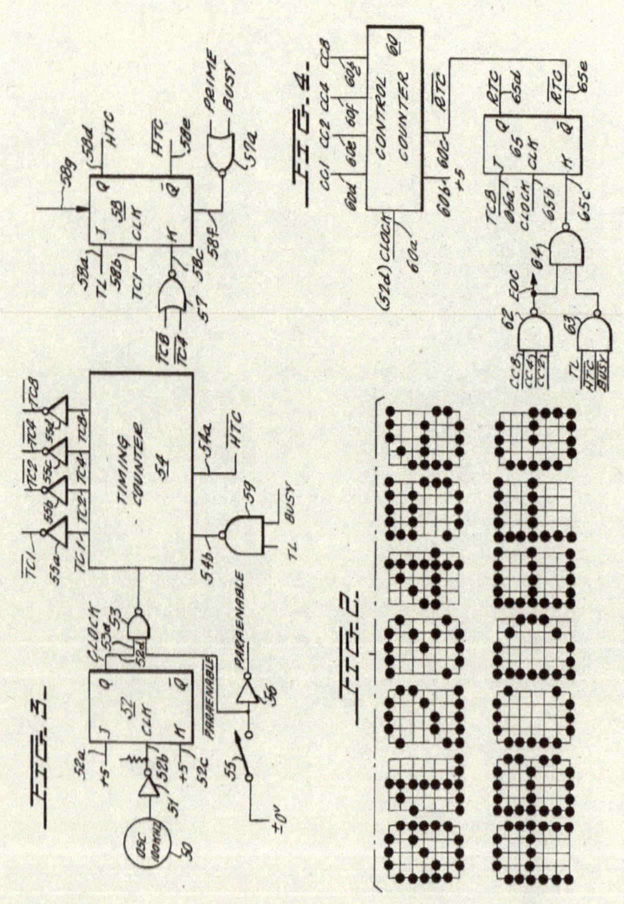
FIG. 3.
FIG. 4.
FIG. 2.
CLOCK
OSC (10 MHz)
50
51
52a
+5
52b
+5
52c
52d
53
53a
Q
52
J
CLR
K
Q
PAREXABLE
55
±0v
56
PAREXABLE
TC1
55a
TC2
55b
TC4
55c
TC8
55d
TC1 TC2 TC4 TC8
TIMING
-COUNTER
54
54a
HTC
54b
TL
59
BUSY
TCB
TC4
TL
58a
TC1
58b
Q
58
J
CLR
K
Q
58d
HTC
58e
HTC
58f
57
58g
57a
PRIME
BUSY
CC1 CC2 CC4 CC8
60d 60e 60f 60g
CONTROL
COUNTER
60
(51d) CLOCK
60a
60b 60c
+5
RTC
TCB
65a
J
Q
65
CLOCK
CLR
K
Q
RTC
65d
RTC
65e
65b
65c
62 EOC
64
63
CC8
CC4
CC2
TL
HTC
BUSY

32

Robert Howard

1923–2014

Dot Matrix Printer
Hudson, New Hampshire
1970

In 2016, Apple removed the headphone jack from the iPhone 7. Owners of the new model were required to use wireless Bluetooth audio or the Lightning port—the only connector on the phone that also charges the battery—for wired headphones. If the headphone jack is a must, owners were encouraged to purchase the Lighting-to-3.5mm audio adapter separately for $9.

The missing headphone jack upset some Apple customers. At the iPhone 7 launch, Apple marketing chief Phil Schiller drove home the company's reasoning, "Maintaining an ancient, single-purpose, analog, big connector doesn't make sense because that space is at a premium." As some tech journalists pointed out, Apple's decision came down to one word: progress.

The 3.5mm headphone jack is based on technology that is more than one hundred twenty-five years old. It is a miniaturized version of the phone connector originally developed in the late 1870s for operators to manually connect telephone calls by plugging cords into a switchboard.

The 3.5mm format was created in the 1950s for the transistor-radio earpiece and was modified in the 1960s for the Sony portable FM radio and again in 1979 for the Sony Walkman. The fact is that the analog headphone jack has been an anachronism since compact disks and other digital technologies like optical audio became available more than thirty years ago.

As with many earlier decisions by Apple—like eliminating floppy disk and CD-DVD drives, replacing parallel ports with USB ports, and adopting Wi-Fi and Bluetooth wireless—the abandonment of the headphone jack, although disruptive, enabled the next generation of technology to develop and flourish.

Interfaces and standards for connecting things together is an important part of electronics and computer history. The adoption of a new format, design, or methodology over earlier ones—like USB for SCSI or Thunderbolt for FireWire—is complex and involves a mix of science, engineering, economics, and a bit of good luck. In some cases, innovation can fill a void and be embraced rapidly if the cost of adoption is affordable. In other instances, persistent obsolescence can override innovation due to design weaknesses or ease-of-use considerations.

Robert Howard—a prolific inventor for seven decades beginning in the 1940s—was among the first engineers to understand that open-technology standards were needed to connect computer equipment together. In the late 1960s, along with Dr. An Wang and Prentice Robinson at Wang Laboratories, Howard developed the thirty-six-pin Centronics parallel interface to connect the Centronics Model 101 dot matrix printer to computers.

Although the Wang Labs team could not have predicted it, the Centronics connector took off and became one of the most successful computer connection technologies ever made. One reason for its success was the performance advantages over previous serial interfaces: parallel could carry multiple data streams between devices and could also simultaneously transmit status information.

More fundamentally, however, was the fact that the computer industry in the 1960s was going through a transition. Prior to the Centronics interface, each computer manufacturer used proprietary solutions designed to block customers from buying equipment from competitors. As the computer peripheral business expanded rapidly, however, the lack of standardized connection methods had become a barrier to progress.

As described by Robert Howard in his autobiography, *Connecting the Dots*, the Centronics parallel port was the beginning of a shift in business philosophy among computer companies, "We came to the conclusion that if we developed a very easy, simple interface and gave it free to the world, it might be accepted and used by everyone. Apparently, the practice of creating unique interfaces was so resented by everyone in the computer industry that once IBM accepted our interface, seven other major companies immediately followed suit." This was not the first or last major technical accomplishment associated with Robert Howard.

*****

Robert Howard was born Robert Emanuel Horowitz in the Brownsville section of Brooklyn, New York, to Samuel and Gertrude (Greenspoon) Horowitz on May 19, 1923. Robert's father worked the midnight shift at the Main US Post Office in New York City. Although he was born three months premature and was afflicted with dyslexia, Robert grew into a very likable and stout youngster with athletic talent in several sports.

After the family moved to Flatbush, Brooklyn, Robert spent much of his spare time at the Brooklyn Ice Palace where he learned to skate. He played youth hockey, and his skills on the ice were noticed by the hockey coach at Brooklyn Technical High School, an elite all-boys public school. Despite his marginal grades, Robert was recruited to attend Brooklyn Tech as along as he promised to improve his studies.

While at Brooklyn Tech, Robert excelled at machine shop and woodworking. He built a model airplane out of balsa wood, tissue paper and a refurbished gas engine as a school project. His 1937 delta-wing design was ahead of its time, and he received an award for it.

Robert was very close to his maternal grandfather, Isaac Greenspoon, who immigrated to the US from Odessa, Russia, in 1910. Isaac started a window-shade business on Manhattan's Lower East Side that became very successful. Robert worked at his grandfather's company as a teenager and acquired business skills and decision-making capabilities that would later prove to be a critical part of his professional success.

Although no one, including family members, expected Robert to graduate, he not only received his high school diploma but was awarded an athletic scholarship to attend the college of engineering at Columbia University. By the time of his graduation from Brooklyn Tech, World War II was well underway, and the Horowitzs changed their name to "Howard" to avoid the anti-Semitism that was on the rise during that period in Europe and America.

Before attending Columbia, Robert took a summer job working the night shift for the Sperry Gyroscope Company in Brooklyn. He was hired to operate the milling and cutting machines used to produce parts for US military searchlights. He kept the job when college classes started so he could cover his living expenses.

In a stroke of good fortune, Robert was hired as an engineer for a new vacuum-tube project at Sperry. Although he was still a student and did not have an engineering degree, the new position required the machine-shop skills that he did have. Robert switched to night school and threw himself into the vacuum-tube development program. This was his first experience with electronics, and like so many other innovators of his generation, the field soon became a focus of his work, and he stuck with it until the end of his career.

After a brief stint in the army, Robert was hired as an engineer at Sylvania Electric Company in Queens, New York. Starting at the age of twenty, he became involved in a seemingly endless series of projects in a wide variety of pursuits that would establish him as a pioneer of postwar electronics innovation. His subsequent accomplishments included the founding of at least twenty-four different companies and the development of dozens of state-of-the-art inventions.

Robert Howard's inventions are so numerous and varied that it is only possible to review a few of them here:

- 1947: Rectangular TV tube

    All early television sets had round picture tubes. This meant that the rectangular broadcast image was either clipped at the top and bottom or was reduced in size to fit in the 7-, 10-, 11-, or 14-inch standard diameters of the first TV tubes. While working for Sylvania, Robert Howard proposed a rectangular tube design and convinced the company to manufacture one hundred of the 16-inch television CRTs.

- 1950: Cable television

    After founding Howard Television, Inc. to build and sell his own design for black-and-white TVs, Howard secured a contract to create the first cable TV system that was designed as part of the newly constructed Windsor Park apartment complex in the Bayside section of Queens, New York. Later known as the master antenna television system (MATS), the project connected eighteen buildings with a total of 320 apartments via coaxial cable to a single television antenna with a signal booster and splitter that enhanced the reception for seven TV channels from the New York area.

- 1961: Improvements in vinyl-record production

    Right around the time that the recording industry was transitioning from 78s to LPs, Howard was collaborating with a company that made the machines that pressed vinyl records. He helped to improve the quality of the mass-produced records by introducing zinc plates into the process. He also invented a pressurized steam-based system for controlling the temperature of the molten vinyl as it was extruded into the record press. Known as the "The Boomer," Robert Howard's invention significantly increased the volume of phonograph-record production while maintaining the highest stereo quality.

- 1968: Casino computer system

    Robert Howard founded Centronics as a division of Wang Laboratories to build the first computerized system to prevent skimming at casino gaming tables. Howard's system tracked the relationship between the amount of cash coming in versus the value of chips going out. The computerized register centrally tracked the amount of each transaction, each table number, and each dealer at any time during the day. Robert Howard's work with the casino industry led to plans for a printing device that could produce multiple hard copy records of gaming transactions. The available technologies of that time were either too expensive and large or too small and slow for this purpose. Working with Doctor Wang at Centronics on a new computer-printing device, Robert's curiosity and sense of entrepreneurship put him on a path of innovation that helped transition the printing industry into the digital age.

- 1970: Dot-matrix printer

Electronic-impact printers with ink-soaked cloth ribbons like typewriters had been developed by IBM in the 1950s for printing from mainframe computers. These machines used a chain with a complete set of characters passing horizontally across the paper at high speed. As the paper moved vertically line by line, type hammers struck from behind and drove the accordion-folded, tractor-fed paper against the ribbon and type characters on the chain. The IBM line of printers had the speed that Robert needed, but they cost about $25,000 and were the size of a large piece of office furniture. While at Wang Labs, Robert developed a self-contained impact printhead which could be made to produce type characters on paper from a matrix of one hundred dots. His invention used wires or "pins" that could print up to 185 characters per second and hit the ribbon and paper hard enough to print all four copies of a multipart form. The core technology of his invention was an electromagnetic switch that could make each pin strike the printing surface one thousand times per second, more than enough to satisfy the performance required for the gaming reports, and at a cost that was affordable. Following the formation of an independent partnership with the Japan-based Brother Industries, Robert Howard's dot-matrix technology was deployed in the Model 101 Centronics printer. Although there were competing dot-matrix devices on the market, Centronics became the most successful mass-production printer of the early computer industry. By the mid-1970s, sales grew exponentially and reached tens of thousands of units internationally. It was the popularity of the printer that made the abovementioned Centronics interface into an industry standard for connecting peripherals to computers that lasted for decades until it was replaced by the Universal Serial Bus (USB) in the 1990s.

- 1991: Direct-imaging press

Robert Howard made what is certainly his most enduring contribution to the printing industry toward the end of his career. In 1986, he founded Presstek to develop the first ever direct-imaging offset printing technology. As he explained in his autobiography, "The problem at that time was that offset color was a slow, costly process. It took at least ten days to two weeks of what was called 'prepress' preparation before a color print job could even be put on a printing press, and because of this expense, it was both impractical and costly to print less than ten thousand copies of anything. I wanted to apply our knowledge of computers and imaging to the color-printing business." Howard's breakthrough concept was to image the printing plates on the press itself and eliminate the darkrooms, film, and chemistry associated with prepress processes. By 1991, a Presstek laser-imaging system was added to a Heidelberg offset printing press and sold as the Heidelberg GTO DI (for direct imaging). At the center of the Presstek system was a set of four-color thermal laser heads that imaged plates on press. Aside from the novelty of the on-press plate imaging, the Presstek technology was waterless and was easily retrofitted onto the existing Heidelberg GTO design because it took the place of the unneeded dampening system. Beginning in 1993, Presstek and Heidelberg developed the Quickmaster DI press, a printing system that was designed from scratch with the on-press laser-imaging technology.

Launched at DRUPA in 1995, the Quickmaster DI became one of the most popular Heidelberg offset presses ever with five thousand machines sold within the decade. The press included design innovations that made it easier to operate than previous offset systems. With this innovation, Robert Howard invented a technology that was both disruptive to the prepress industry and enabled former prepress companies to enter the short-run color-printing market.

Robert Howard died on December 19, 2014, at the age of ninety-one. Although he is not a well-known figure in the history of printing—perhaps because of the variety of businesses and disciplines where he left his mark—Robert made critical contributions to the industry, especially in the final decades of the twentieth century. His exceptional talents as an engineer and entrepreneur were essential to the transition of offset printing from an exclusively analog process to one that uses a host of integrated digital technologies.

33

Efraim "Efi" Arazi

1937–2013

Color Electronic Prepress System
Tel Aviv, Israel
1979

One of the most important achievements of personal computers and mobile wireless technologies is that they have made it possible for the public to do things that could previously be done only by professionals. Take video, for example. It is easy to take for granted the video-production functions that are performed routinely today on inexpensive and easy-to-use mobile devices. In the early 2000s, the ability to capture and edit HD video would have cost tens of thousands of dollars in digital camera and production equipment and required extensive training to use it.

The same can be said for the ability to quickly create a document in a word-processing program and insert high resolution graphics anywhere on the page, cropping and scaling as needed. Applying filters and adjusting image quality (contrast, brightness, sharpness) is also second nature as these functions are today available on every mobile device.

In the 1970s, before the rise of personal computers, electronic image editing, scaling, and cropping could only be performed on very expensive prepress systems that cost more than one million dollars. This was during the era of what was known as color electronic prepress systems (CEPS) built on state-of-the-art minicomputers with reel-to-reel magnetic tape for data storage.

During the 1960s and 1970s, as commercial offset lithography and film-based color reproduction were overtaking letterpress and single-color work, high-end digital electronic production systems were acquired by big printing companies and major magazine and newspaper publishers that could afford to make the investment.

By the 1960s—after analog electronic systems had been widely adopted in pressrooms and prepress and typesetting departments across both Europe and America—a race was on to develop a computerized page-composing system. Companies like Hell, Crosfield, Dai Nippon Screen, 3M, and others that had gone through the postwar electronics revolution jumped into the market to try and solve the problem of merging text and color photographs together electronically on a computer display.

However, it was a newcomer to the graphic-arts industry from Israel called Scitex, founded by Efraim "Efi" Arazi in 1968, that made the highly anticipated breakthrough. Foreshadowing the impact of PC-based desktop publishing on graphic communications in the mid-1980s, Scitex introduced digital files and computerization to the prepress production process and forever changed the printing industry.

Efi Arazi was born in Jerusalem on April 14, 1937 and entered the Israeli military at sixteen and a half years old and without graduating from high school. He made a name for himself as an exceptional electronics specialist while working on radar systems in the Israeli air force. With the assistance of the US embassy, Arazi was admitted to the Massachusetts Institute of Technology (MIT) in 1958 as an "extraordinary case" despite his lack of the normally requisite secondary-school diploma.

While attending MIT, Arazi also worked at Harvard University's observatory and digital photography lab. Under the direction of Harvard Professor Mario Grossi, Arazi petitioned NASA and was awarded funds to develop a camera system that was used on the unmanned lunar probes in 1966 and 1967 to scan the surface of the moon. It has also been reported that Arazi's invention was part of the equipment on the Apollo 11 mission that captured and transmitted video to earth of Neil Armstrong's first footsteps on the Moon on July 20, 1969.

After earning a bachelor's degree in engineering at MIT, Arazi worked in the US for a short time for Itek corporation, a US-defense contractor that specialized in spy satellite imagery. In 1967 he returned to Israel, and one year later—along with several others who had been educated in the US—founded Scientific Technologies (later shortened to Scitex) with the aim of developing electro-optical devices for commercial purposes.

Scitex's first products were developed for the textile industry. The company sold nearly one hundred electronic systems that automated the process of creating knitting patterns. Since many colors were used in complex fabric designs, such as the popular Jacquard pattern, Arazi and Scitex developed a scanner (Chroma-Scan) and image-manipulation workstation (Response 80) that programmed electronic double knit-stitching looms.

These optical systems replaced manual and time-consuming stitch-by-stitch drawings and punch cards that were used widely in the textile industry up to that time. Scitex also later devised a system for imaging film for printing on textiles that included overprinting, trapping, and repeating patterns. In the early to mid-1970s, Scitex also brought raster imaging and CRT technologies to the cartography, seismography, and printed circuit-board industries. These were secondary markets for Efi Arazi and his Scitex team.

Recognizing the potential for new technologies in the growing international printing and publishing markets, Scitex began development in 1975 of a computerized color prepress system. Arazi stunned the international graphic arts industry in the fall of 1979 when he demonstrated the Response 300 system for the first time at the GEC expo in Milan, Italy.

Response 300 included an integrated color drum scanner, image-editing workstation, and laser film plotter. Directly challenging the domination of high-tech graphic arts equipment by Hell (Germany) and Crosfield (UK), Scitex was the first company in the world to combine color image retouching and page makeup onto a single console.

Prior to the Response 300, the electronic color-scanning process was based on an analog transfer of color-separation information directly from a drum scanner to the film output device. The innovation of Arazi and Scitex was to place a minicomputer—at that time an HP1000—between the scanner and plotter such that the color separations were captured and stored in digital form. The proprietary image files could then be color corrected, retouched, scaled, and cropped on screen prior final output to film separations.

In describing the significance of the accomplishment, industry historian Andy Tribute later explained, "It allowed you to do in real time on a terminal the sort of things we do in Photoshop now.… I remember watching Efi do a demo where he had a picture of a person with a Rolex watch on, and he changed the date in real time on the Rolex. Today that may seem nothing, but back then, it blew my mind."

Within one year, Scitex had sold one hundred million dollars of the Response systems to printers and publishers. Through the mid-1980s, Arazi led Scitex as it developed a suite of products (Raystar, SmartScanner, Whipser, Prisma and Prismax Superstation to name a few) that brought the latest in minicomputer technologies to high-end prepress workflows. Scitex customers gladly paid the one-million-dollar price tag for the flexibility and time savings that these CEPS systems provided.

Scitex remained an innovator throughout the 1980s and 1990s as proprietary technologies, and CEPS gave way to desktop publishing, industry-standard file formats, and PostScript workflows. Scitex was among the first prepress technology companies to embrace the introduction of Macintosh computers into graphic-arts production.

In 1988, Scitex partnered with Quark technologies, developer of the most sophisticated desktop publishing software at the time and made it possible for QuarkXPress users to build compound documents with high-resolution, full-color images to be output in commercial- and publication-printing workflows.

In 1985, Arazi pushed the industry forward with the development of Handshake, a Scitex product that allowed a wide variety of systems including those of competitors to send and receive data from the Response line of products. Later, Scitex was an advocate of Digital Data Exchange Standards along with Hell, Crosfield, Eikonix, and others to smooth that transfer of data between all systems in the industry.

In June 1988, Arazi stepped down as president and CEO of Scitex. Six months later, when Mirror Group's Robert Maxwell acquired a controlling stake in Scitex, Efi Arazi also resigned as chairman of the board. While the company had reached the height of its success with revenues approaching one billion dollars and four thousand employees, Arazi knew that personal computers were transforming the industry, and it was time to move on to other business ventures.

After Arazi's departure, Scitex continued to develop prepress-workflow systems, laser-imaging equipment, desktop scanners, digital color, and soft-proofing devices. The company participated in the transition from film-based workflows to the direct-to-plate revolution of the 1990s.

Along with all its competitors, Scitex began to struggle financially and ended up, after several ownership transitions, selling its graphic arts group to Vancouver-based competitor Creo Products in 2000. The division of the company that went into digital printing called Scitex Vision was acquired along with the Scitex name by HP in 2005, and the remainder of the business was renamed Scailex at that time.

In 1988, Efi Arazi founded Electronics for Imaging (EFI) at the age of fifty-one. The new venture was no less successful than Scitex as EFI raster-image processors were integrated in many high-quality color laser and toner-based printing devices. The EFI Fiery technology became a standard in the graphic-arts industry by the 1990s. The company—which bears the first name of its founder as an acronym—later expanded into ink-jet printing devices, printing-industry productivity software, and print server and workflow software tools. Efraim Arazi died on April 14, 2013, at age seventy-six.

34

Steve Jobs
1955–2011

Desktop Publishing
Cupertino, California
1985

Steven Paul Jobs's role in creating one of the first personal computers along with his neighborhood friend Steve Wozniak, the founding of Apple Computer and his subsequent firing and return to the company, have become part of tech-industry lore. His later contributions to mobile, wireless, and touch computing—embodied in the Apple iPod, iPhone, and iPad—were no less transformative.

Although Steve Jobs had extensive knowledge of computer hardware, operating systems, and applications—he even worked for a short time in the early 1970s as a technician for Atari—his greatest skills were as technology visionary, marketer, and salesman. Without the entrepreneurial drive, leadership charisma, and design aesthetic of Steve Jobs, Apple would never have emerged as the world's largest publicly traded corporation or have the most loyal customers in the history of the consumer-product industries.

Owing a great deal to the location and times of his upbringing—Silicon Valley in the 1950s and 1960s—Steve Jobs expressed a broad cultural viewpoint and considered every project and product an aspect of a larger creative purpose. Having developed an enthusiasm for the Bauhaus movement's form and function philosophy, he identified design simplicity with products that were both beautiful and easy to use.

In his 2012 biography of Steve Jobs, Walter Isaacson quotes the Apple cofounder from a presentation he gave in the 1980s, "So that's our approach. Very simple, and we're really shooting for Museum of Modern Art quality. The way we're running the company, the product design, the advertising, it all comes down to this: Let's make it simple. Really simple." Jobs rejected the boxy, bulky, and dark industrial style of the earlier generation of computer design in favor of elegance and what he later called "taste."

It was out of this unique blending of art with science and business that Steve Jobs made two significant contributions to typography and printing technology: the creation of the first fonts on a personal computer and the launching of desktop publishing. As with every innovation associated with his name, Jobs relied on the skills of others to realize his vision and then packaged and presented the accomplishments with great fanfare to investors and consumers alike.

Jobs's aesthetic sensibility had been formed a decade earlier while he was briefly a student at Reed College in Portland, Oregon, in 1972. After dropping out of school, he enrolled in a calligraphy course at Reed taught by Father Robert Palladino. The course had a lasting impact on him.

As Jobs explained in a commencement address he delivered to Stanford University in 2005, "Reed College at that time offered perhaps the best calligraphy instruction in the country.... I learned about serif and sans serif typefaces, about varying the amount of space between different letter combinations, about what makes great typography great. It was beautiful, historical, artistically subtle in a way that science can't capture, and I found it fascinating.

"None of this had even a hope of any practical application in my life. But ten years later, when we were designing the first Macintosh computer, it all came back to me. And we designed it all into the Mac. It was the first computer with beautiful typography."

While working with the Macintosh design team, Jobs was involved in every detail of the computer's size, shape, and color as well as every icon, window, and menu of the graphical user interface. This involvement included the creation of fonts which he insisted be named for the great cities of the world: Cairo, Chicago, Geneva, London, Los Angeles, Monaco (mono-spaced system font), New York, San Francisco, Toronto, and Venice.

Prior to the work of the Apple team—including designer Susan Kare, developer Andy Hertzfeld and engineer Bill Atkinson—on proportional fonts, computers were mostly limited to mono-spaced typefaces much like a typewriter with all alphanumeric characters and keystrokes the exact same width. Jobs could see that the bit-mapped display of the Macintosh desktop was capable of rendering typefaces with a sophistication which approximated that of letter-press hot-metal type and cold phototypesetting.

Others at Apple Computer, due to their limited perspective on the utility of the personal computer, could not relate to Steve Jobs's insistence on the font library; they considered it a distracting personal obsession. In his biography of Jobs, Walter Isaacson quotes Apple investor and partner Mark Markkula: "I kept saying, 'Fonts? Don't we have more important things to do?'"

When Steve Jobs launched the Macintosh on January 24, 1984, at the Flint Center in Cupertino, the font library was a critical part of the presentation of "the computer for the rest of us." It was the first desktop system to offer not only the nine city-named fonts listed above but also style choices—Plain, Bold, Italic, Bold Italic, Underline, Outline, Shadowed—for each.

While initially appearing somewhat primitive, bitmapped, and lacking the finesse of professional typography, Jobs's on-screen fonts were the beginning of a revolution in type technology. For the first time in print media history, fonts became something that everyone with a computer could use, not just professional graphic designers and industry specialists.

The Macintosh font library also encouraged professionals to push the limits of computer-generated typography and eventually transformed the field of typesetting altogether. Soon, desktop fonts surpassed the quality and versatility of all previous type technologies and offered WYSIWYG (What You See Is What You Get) output—

that is, the image displayed on the computer screen is precisely what is printed onto a sheet of paper or other final output media.

Steve Jobs understood the promise of WYSIWYG long before the phrase was widely used in the printing and publishing industries. Nearly one year to the day after unveiling the Macintosh, Jobs was back on stage in Cupertino at the annual Apple stockholders meeting on January 23, 1985, to launch the Apple LaserWriter and demonstrate the first ever desktop publishing system.

Desktop publishing signifies an integrated publishing system whereby pages containing both text and graphics are designed in layout software on a desktop computer and printed in individual or multiple copies on a desktop printer. Building on the accomplishments of the Macintosh, Steve Jobs worked throughout 1984 with partner companies and publishing-industry experts to integrate the Apple Macintosh computer with other basic elements of desktop publishing: the Apple LaserWriter, Adobe PostScript, and Aldus PageMaker.

- Apple LaserWriter: Gary Starkweather invented the core toner imaging technology of the laser printer at Palo Alto Research Center (PARC) in the early 1970s. Although Xerox never brought a desktop laser printer to market, HP and Canon developed systems independently of each other in the 1970s. The HP LaserJet, based on the Canon LBP-CX printing engine, was the first desktop laser printer and was released in 1984. The Apple LaserWriter, developed by a team led by project manager Bruce Blumberg, had two important differences with the HP device: it was networked (with AppleTalk) and could be shared and contained breakthrough PostScript software that enabled true WYSIWYG output. The Apple LaserWriter was available for purchase in March 1985 and sold for $6,995.
- Adobe PostScript: The software at the heart of the Apple LaserWriter was Adobe's PostScript page-description language. John Warnock and Chuck Geschke, who also came from Xerox PARC, founded Adobe Systems in 1982 with PostScript as their flagship product. Warnock and Geschke

developed a state-of-the-art device independent print-programming language that (1) captured all the elements—text, graphics, geometry, etc.—on the page of the desktop layout software during the "Print" function; (2) interpreted the layout data as vector-based objects within the memory of the printer; and (3) converted the PostScript objects into raster-print data such that the printer could render it onto a sheet of paper at a resolution of three hundred dots per inch. The Adobe founders also signed a licensing agreement with Linotype Corporation that made thirteen professional typefaces (four styles for each of the Helvetica, Times New Roman, and Courier families, plus a Symbol font) "resident" within the PostScript raster image processor (RIP) in the Apple LaserWriter.

- Aldus PageMaker: Paul Brainerd—the man who coined the phrase "desktop publishing"—cofounded Aldus Corporation in February 1984 in Seattle, Washington. With a background in computerized newspaper publishing systems, Brainerd and a group of developers began working on layout software initially intended for the newspaper market. After getting some early peeks at the Apple Macintosh, Adobe PostScript, and Apple LaserWriter, the Aldus team developed PageMaker as the first application capable of placing columns of text and images onto a virtual page and used a floating tool palette. The first commercially available version of PageMaker was released in July 1985 and sold for $495.

An important advisor to Steve Jobs throughout the launching of the first desktop system was John W. Seybold, a pioneer in computerized publishing systems and industry consultant. According to Paul Brainerd, "There were a couple of people that really were the glue that made all this come together, and the most important was Jonathan Seybold. He was consulting to both Adobe and Apple. He and I knew each other for a long time going back…he told me some time during '84, probably in the first quarter, that there was some confidential information that I needed to know. He got clearance from his clients to be able to share it with me."

In an account published by Adobe in 2004, John W. Seybold reviewed the significance of the events that unfolded during the summer of 1984, "Steve wanted to see me urgently. He said they had a deal with Adobe, they were signing a deal with Linotype, they had real fonts. I went to Cupertino and walked into this tiny room, and there stood Jobs and Warnock with a Mac and a LaserWriter. He showed me what they were up to. I turned to Steve and said, 'You've just turned publishing on its head. This is the watershed event.'"

Although they are less celebrated, Steve Jobs's introduction of the Apple Macintosh font library and his pivotal role in launching the desktop publishing revolution in 1984–1985 were disruptive innovations on a par with the invention of printing because they made designing and publishing accessible to anyone with a personal computer and printer. The lasting impact of Jobs's breakthrough continues to be felt today in the form of online and social media publishing by billions of people across the globe. Jobs's death from cancer at age fifty-six on October 5, 2011, prematurely ended the life of one of the most significant and creative figures of our times.

Appendices

Appendix 1

Two Millennia of Paper

We often take paper for granted. When searching for a dollar bill, filling up a fountain drink cup, or moving a leaf bag to the curb, do we think about paper? Probably not. We are focused on the useful purpose of these daily items and do not have time to stop and think about how they are made or what they are made of. Paper in all its different forms, qualities, and applications has been around for a very long time. Most commonly, paper is thought of as a medium for the written or printed word. This is natural since paper—the word is derived from the Latin term *papyrus*—was developed as a writing surface nearly two centuries ago by the Chinese to replace wood and bamboo scrolls.

The papermaking process—basically unchanged since Ts'ai Lun (Cai Lun) invented it in AD 105—is a marvel of human ingenuity. Distinct from the papyrus of ancient Egypt, where thinly cut plant stalks were woven and laminated together, paper is the reduction of a raw material to individual fibers and their liquid suspension onto a mat or sheet until dry.

With today's world of instant global communications and international travel, it might seem strange that it took five hundred years for papermaking to leave China and arrive in Japan and nearly one thousand years for it to reach Europe. Nonetheless, the story of paper's global expansion and development is an important chapter of world history.

In the seventh century, the Japanese were the first to recycle and repulp paper. In AD 750, after a battle between the Chinese and the Muslims in what is now Uzbekistan, a group of Chinese prisoners revealed their secrets to their Middle Eastern captors. Once the Muslims began making paper, they went on to develop water-powered stamping/hammer mills for the pulping process.

Papermaking entered Europe through the Muslim Moors of southern Spain in about AD 1,100. At that time, most European documents were recorded on parchment, a writing surface of sheepskin or vellum from calfskin. Since Europe was majority Christian and it was the time of the Crusades, the papermaking techniques of the Moors were not discovered by Europeans until after the military campaigns were concluded in the south. Once the Vatican was exposed to the wonders of papermaking, Italy emerged as the primary producer of paper in Europe.

In the thirteenth and fourteenth centuries, the center of papermaking moved from Italy to France when the craft was encouraged by the monarchy. Just as demand for handwritten documents was on the rise, metallurgist Johann Gutenberg invented the mechanical methods of metal typesetting and printing in Mainz, Germany, in 1440. As the printing press spread throughout Europe, the volumes of paper being produced by the mills in France and Italy grew exponentially.

By the eighteenth century, papermaking had moved largely to Germany and Holland due to the social and political instability in France. Meanwhile, the technology of the paper industry was undergoing a transformation brought on by the emergence of manufacturing throughout Europe. These are the same industrial developments that impacted printing presses: iron in place of wood, steam in place of manpower or other natural forces such as wind and water.

In 1800, a Frenchman named Nicholas-Louis Robert patented an invention that converted papermaking into a mass-production industry. Robert's paper machine had a continuous wire screen upon which the slurry was poured so that the excess water would pass straight through it. The paper in formation was progressively dried by a series of felt rollers until it was solid enough to be wound onto a roll. Thus, paper no longer needed to be made in individual sheets.

Several years later, Robert's invention was sold to the Fourdrinier brothers of London where they constructed a much larger version of it. In 1812, the first Fourdrinier—the name associated with Robert's invention and remains the primary method for papermaking to this day—machine was started up in a mill near Two Waters, England. Later, cylinders for pulp transport, drums for drying, and techniques to prevent ink absorption into the fibers of the paper (sizing) would modify the Fourdrinier system.

By the mid-nineteenth century, the center of papermaking moved to America and played an important role in the growth of newspaper publishing around the time of the Civil War. Up to this point, the fiber for papermaking, especially in Europe, came from the fabric in rags. However, with the vast forests of North America, wood fiber quickly became the source of paper pulp and ground-wood the essential raw material for the newsprint industry.

In the twentieth century, as printing technology moved from black-and-white letterpress to full color offset lithography, coated papers were developed. The papermaking process evolved from offline to in-line coating systems. Today, the pulp-and-paper industries worldwide are going through a transformation born of the global economy and the shifting of paper consumption from west to east. According to industry data, paper consumption in the advanced world is falling rapidly—brought on by electronic media and recycling practices—while paper consumption in the developing world is rising even more rapidly. In 2009, for example, paper consumption in China surpassed that of the United States for the first time.

While paper remains the number one media for publishing, electronic and online alternatives have been in development and grown rapidly over the last several decades. In the 1970s, Nick Sheridon at Xerox's Palo Alto Research Center (PARC) developed the world's first electronic paper called Gyricon. It consisted of microscopic polyethylene spheres with black on one side and white on the other embedded in a silicon sheet. With the appropriate electronic charge, this e-paper could be used repeatedly to display an unlimited number of different images much like a computer monitor.

In 2007, Amazon began marketing the Kindle e-book reader with a paper-like display, based on a principle similar to Sheridon's invention. The Kindle emulates the visual characteristics of book paper because it relies upon reflective light as opposed to the transmissive backlighting of computer displays. Although these technologies lack the surface flexibility of paper, there are developments underway that will soon bring that attribute to electronic publishing. For example, in March 2012, LG unveiled the world's first commercially available six-inch e-paper display that can be bent at an angle of up to 40 degrees and foldable displays will soon follow.

While good old-fashioned paper will be around for a long time—ensured by its utility, durability, recyclability, and cost—the one sector where we can now visualize its decline and possible disappearance is in the publication of books, magazines, and newspapers. If and when this happens, perhaps we will better appreciate the miracle of paper and no longer take it for granted.

Appendix 2

The Book Called *The Book*

The flattening of e-book sales has been underway for years and comes as no surprise. The exponential growth rates of the e-book market that began in 2008—peaking at 254 percent in the first quarter of 2010—could not be sustained indefinitely. As Harper-Collins CEO Brian Murray noted in November 2013 to *Publishers Weekly*, "Nothing grows by triple digits for too long."

The inflection point in the e-book market is an appropriate moment to take note of the publication in 2014 of the volume *The Book: A Global History* by Oxford University Press. Edited by Michael F. Suarez and H. R. Woudhuysen, *The Book* contains fifty-four essays in two sections: Thematic Studies (twenty-one essays) that deal with the origins of written communications, the ancient history of books, and the technologies of book production; and Regional and National Histories of the Book (thirty-three essays) that cover book history in many countries in Europe, Asia, Australia, and the Americas.

The new volume is a concise edition of the highly acclaimed *The Oxford Companion to the Book* published in 2010. That work by the same editors is 1,327 pages, comes in two vivid burgundy-bound volumes with gold lettering, costs $325, and is clearly intended as library-reference material.

As Suarez and Woudhuysen state in the introduction to the new smaller volume, "Encouraged by the commercial and critical fortunes of *The Oxford Companion to the Book* (2010), its general editors were nonetheless chastened by the fact that its costliness prevented many from purchasing the two-volume work of 1.1 million words.... It is our hope that the publication of this volume will make a valuable collection of bibliographical and book-historical scholarship—ambitious in its scope and innovative in its reach—accessible to a broad audience of general readers and advanced specialists alike."

One can tell from these two sentences that this is not a book to be read for entertainment purposes. In any case, *The Book: A Global History*—despite its pedanticism and academic erudition—is an important resource. The research and scholarship enable those in modern-day print communications to gain a deeper understanding of their profession and its social, technical, and economic origins and development over many centuries.

For example, in a significant point in the introduction, the editors describe the term "book" as a synecdoche—that is, a figure of speech that represents all other textual communications forms. They write, "Naturally, the use of the term 'book' in our title in no way excludes newspapers, prints, sheet music, maps, or manuscripts…in every European language, the word for 'book' is traceable to the word for 'bark,' we might profitably think of 'book' as originally signifying the surface on which any text is written and, hence, as fitting shorthand for all recorded texts."

It is, of course, not possible to summarize the entirety of the new volume here, so below are a few highlights followed by some concluding thoughts:

- Writing Systems

    The very first essay by Andrew Robinson is about the origin of human writing systems. It should come as no surprise that the birth of writing is thought to have taken place in what is now known as the "cradle of civilization" in Mesopotamia sometime in the late fourth millennium BC. The impulse for the first writing system was the need for record keeping associated with trade. As Andrews writes, "The complexity of trade and administration had reached a point where it outstripped the power of memory among the governing elite. To record transactions in an indisputable, permanent form became essential." One theory holds that the system began with the exchange of clay tokens and that this was later substituted with two-dimensional symbols written in clay. Of course, another origin of writing is the Ice Age symbols found in caves in southern France that date back twenty thousand years. These images, however, are considered "proto-writing" because they do not conform to the scientific definition of writing provided by the noted Sinologist John DeFrancis as a "system of graphic symbols that can be used to convey any and all thought."

- The Book as Symbol

   In essay number 7, Brian Cummings treats the book as an object within different historical contexts. He writes, "A book is a physical object, yet it also signifies something abstract, the words and the meanings collected within it. Thus, a book is both less and more than its contents alone." During ancient times, books were considered sacred, and access was restricted to a priestly class. Writing was itself considered a spiritual power and consumption of the written words also a mystical act. Later, Plato posited that writing was inferior and untrustworthy in comparison to the knowledge within the mind. In biblical times, "the scrolls" (Torah), "the tablets" (Ten Commandments), and "the scriptures" (the Bible) emerged as those writings that embodied divine power, law, and ordinance. Cummings writes, "The incorporation of the holy book into ritual—the raising of the book or carrying of it in procession or kissing of the book or kneeling before it—is only the most obvious example of this." These practices exist today in the courtroom or inaugural "swearing in" and "taking the oath" of witnesses and officials who place their hand upon the Bible. Another more modern and historically significant treatment of books as symbols was the infamous Nazi burning of "degenerate" books in Berlin's Opernplatz in 1933.

- The Electronic Book

    The essay by Eileen Gardiner and Ronald G. Musto on the e-book was revised and expanded from the *Oxford Companion to the Book* of 2010, and this fact expresses the rapid development of the new form. Of note is the essay's summary of the background to present-day e-books. In 1945, Vannevar Bush provided the earliest documented vision for proto-electronic books in what he called the Memex. This hypothetical system would store the entire contents of scientific literature and had pages, page-turners, annotation capability, internal and external links, storage, retrieval and transmission of documents. But while Bush conceived of the Memex on microfilm, the first truly electronic book concept came in 1965 with the "hyperlinking," "hypertext," and "hypermedia" vision of Ted Nelson. Nelson's idea was first demonstrated—now famously referred to as "The Mother of All Demos"—by Douglas Engelbart. The system was called NLS (for oN-Line System) and was developed by the Stanford Research Institute. Engelbart demonstrated in combination for the first time anywhere the following: email, teleconferencing, videoconferencing, and the mouse. The future of hypertext from this point forward merges directly with the emergence of the personal computer and later the World Wide Web—that is, Apple HyperCard by Bill Atkinson in 1987, the emergence of CD-ROM books in 1992, online and searchable texts accessible with the Mosaic/Netscape web browser in 1993.

As mentioned earlier, *The Book: A Global History* is an important resource for bibliophiles and anyone seeking thorough knowledge of the history of print communications. Going through the volume, readers gain a better understanding of world history. It seems that books, both their antecedents and descendants, are intermingled with the history of all civilization from the end of human prehistory to the present.

With the development and adoption of e-books—combined with wireless and mobile communications—it is possible to envision a future where print-on-paper books are recognized as relics of a distant past. But such an eventuality would also require a development well beyond the present e-book reader and mobile media experience along with solving many other complex problems. While great strides have been made, we are still at the very beginning of the transition from the completely analog world of books to the completely digital world of instantaneous and infinite access to all information, knowledge, and culture.

Appendix 3

The Birth, Life, and Afterlife
of Newspapers

The year 2014 marks twenty-five years of the World Wide Web. If you search online for "the birth of the web," you will find a link to the page from the website of CERN (European Organization for Nuclear Research) devoted to the Web's invention.

In March 1989, software engineer and networking specialist Tim Berners-Lee submitted a proposal at CERN for an online system of information sharing among ten thousand scientists in over one hundred countries. His basic concept was to merge personal computers, networks, and hypertext into a global information system. Originally calling the system "Mesh," the proposal refers to a "web" of documents where a "hypertext browser" could be used to view data and information.

There are many important facts to be found on the CERN site about the creation of the Web by the thirty-three-year-old English-born Berners-Lee and how his idea produced the most important media technology innovation of our lifetime. If you are inclined, you can view Berners-Lee's original draft proposal in HTML, text, or as a PDF of the printed MacWord document that was distributed at CERN.

Although it is commonplace today, our ability to view and read these documents online from anywhere at any time is a product of the very transformation that they brought about. But the Web is not only the repository of viewable documents from the moment of the

Web's invention forward; it is also becoming the online warehouse of documents that previously existed only in printed form.

Among the expanding digital library of printed archives is the vast universe of the world's newspapers. Next to books, newspapers are the second most important paper-based knowledge and information source. And, due to their monthly, weekly, or daily frequency, newspapers are the only existing chronological record of so many accomplishments, endeavors, hobbies, errors, and catastrophes of people on a local, national, and global scale.

Newspapers as a form of print communications began in the early seventeenth century. The first newspapers were circulated in Germany, and the expansion of the form is closely associated with the spread of the printing press (hence the term "press" to describe these publications). It is said that the *Strasbourg Relation*, printed by Johann Carolus beginning in 1609, was the first newspaper to appear because it had a publishing frequency and contained a variety of news items.

In the 1600s, newspapers spread rapidly throughout Europe (Holland, England, France, Portugal, Spain, Sweden, and Poland) and to the American colonies. In 1690, Benjamin Harris published the first American newspaper in Boston called *Publick Occurrences Both Forreign and Domestick*. It lasted for one issue before colonial authorities shut it down.

Alongside political pamphlets, such as Thomas Paine's spectacularly popular *Common Sense*, American newspapers played a critical role in the revolution of 1776. Benjamin Franklin was a leading publisher—having financed numerous printing establishments throughout the colonies—and worked on a plan for an intercolonial network of newspapers.

The era of the dramatic rise of newspapers began with the industrial revolution; publications exploded across the United States throughout the 1800s. Paralleling the westward expansion and the rapid population increase across the country, the number of newspapers shot from 329 in 1800 up to 15,872 by 1900 and reached a peak at 17,083 in 1910. The number of daily newspapers in the US also reached an apex in the same time frame, with more than 2,500 by the end of the first decade of the twentieth century.

While the communications technology advancements of the 1800s—innovations in printing, typesetting, photographic reproduction, and distribution as well as the telegraph—accelerated the process and reduced the cost of print publishing, significant news media alternatives to newspapers emerged in the twentieth century. First commercial radio in the 1920s and then television in the 1960s augmented the newspaper business.

In our time, the Internet and the World Wide Web have had a similar, albeit more dramatic, impact on newspapers. However, newspapers continue to exist, and although daily papers have been falling steadily since 1980, the total number of newspapers in the US has been flat since 1990 and is greater today than it was in 1960.

The point here is that most of the two hundred years' worth of newspaper pages lie in physical archives, some well-maintained while others are deteriorating by the hour. Efforts to preserve printed newspapers by converting them to searchable database of scanned images and OCR (optical character recognition) text have been mounted although not without challenges and difficulties.

For example, Google launched with great fanfare a news archive of the world's newspapers on June 6, 2006. This initially turned out to be the database of PaperofRecord.com, a firm that had been acquired by Google. But later, on September 8, 2008, Google announced the expansion of the concept with the launch of an indexed database of scanned newspapers dating back to the 1800s. This project scanned some two thousand archives.

But on August 14, 2011, without notice or explanation, Google canceled its news archive service. Having previously gone through a lengthy and debilitating legal fight over the Google Books program, it appeared that the search giant had been bullied by the prospect of a long, drawn-out copyright battle with newspaper publishers over copyright issues.

As of December 2013, the Google news archive of more than two thousand of the world's most important newspapers has become accessible and searchable at http://news.google.com/newspapers.

One of the advantages of the Google archive is that it is open and freely searchable. However, initiatives to create free access newspaper archives—such as the Library of Congress at http://chroniclingamerica.loc.gov where 6.6 million newspaper pages from 1836 to 1922 have been made available to the public—are overwhelmed by subscription and fee-based online sources. Some systems offer sophisticated search tools combined with subscription and/or membership fees.

A comprehensive listing of both the "free and pay wall blocked" digital online newspaper archives has been published on Wikipedia: http://en.wikipedia.org/wiki/Wikipedia:List_of_online_newspaper_archives.

Browsing the Google news archive, it is interesting to look back and read how the World Wide Web was viewed around the time that it was developed. An article "Once obscure, the World Wide Web is mainstream" appeared on page 17 of the *Reading Eagle* of Reading, Pennsylvania, on July 28, 1996. With 12 percent of the US population over age sixteen at that time accessing the Web, Associated Press journalist Evan Ramstad noted, "The Web in a decade or so may be so common that people won't think about it. Like the phone system on which it relies, the Web will just be there."

Fortunately, the newspaper archives hosted on the Web have also made it possible for us to find and read today what was being said and written about the Web while it was being created.

Appendix 4

Genesis of the Graphical User Interface

Thirty-five years ago, Xerox made an important TV commercial. An office employee arrives at work and sits down at his desk while a voice-over says, "You come into your office, grab a cup of coffee, and a Xerox machine presents your morning mail on a screen.… Push a button, and the words and images you see on the screen appear on paper.… Push another button, and the information is sent electronically to similar units around the corner or around the world."

The speaker goes on, "This is an experimental office system. It's in use now at the Xerox research center in Palo Alto, California." Although it was not named, the computer system being shown was called the Xerox Alto, and the TV commercial was the first time anyone outside of a few scientists had seen a personal computer.

The Alto is today considered among the most important breakthroughs in PC history. This is not only because it was the first computer to integrate the mouse, email, desktop printing, and Ethernet networking into one computer; above all, it is because the Alto was the first computer to incorporate the desktop metaphor of "point and click" applications, documents, and folders known as the graphical user interface (GUI).

The real significance of the GUI achievement was that the Xerox engineers at the Palo Alto Research Center (PARC) made it possible for the computer to be brought out of the science lab and into the office and the home. With the Alto—the hardware was conceptualized by Butler Lampson and designed by Chuck Thacker at PARC in 1972—computing no longer required arcane command line entries or text-based programming skills.

The public could use the Alto because it was based on easy-to-understand manipulation of graphical icons, windows, and other objects on the display. This advance was no accident. Led by Alan Kay, inventor of object-oriented programming and expert in human-computer interaction (HCI), the Alto team set out from the beginning to make a computer that was "as easy to use as a pencil and piece of paper."

Basing themselves on the foundational computer work of Ivan Sutherland (SketchPad) and Douglas Engelbart (oN-Line System), the educational theories of Marvin Minsky and Seymour Papert (Logo), and the media philosophy of Marshall McLuhan, Kay's team designed an HCI that could be easily learned by children. In fact, much of the PARC team's research was based on observing students as young as six years old interacting with the Alto as both users and programmers.

The invention of the GUI required two important technical innovations at PARC:

- Bitmap computer display

  The Alto monitor was vertical instead of horizontal, and with a resolution of 606 by 808 pixels, it was 8 × 10 inches tall. It had dark pixels on a light-gray background and, therefore, emulated a sheet of letter-size white paper. It had bitmapped raster scan as a display method as opposed to the "character generators" of previous monitors that could only render alphanumeric characters in one size and style and were often green letters on a black background. With each dot on its display corresponding to one bit of memory, the Alto monitor technology was very advanced for its time. It was capable of displaying multiple fonts and could even render black-and-white motion video.

- Software that supported graphics

  Alan Kay's team developed the SmallTalk programming language as the first object-oriented software environment. They built the first GUI with windows that could be moved around and resized and icons that represented different types of objects in the system. Programmers and designers on Kay's team—especially Dan Ingalls and David C. Smith—developed bitmap graphics software that enabled computer users to click on icons, dialogue boxes, and drop-down menus on the desktop. These functions represented the means of interaction with documents, applications, printers, and folders and thereby the user-derived immediate feedback from their actions.

The Alto remained an experimental system until the end of the 1970s with two thousand units made and used at PARC and by a wider group of research scientists across the country. It is an irony of computer and business history that the commercial product that was inspired by the Alto—the Xerox 8010 Information System or Star workstation—was launched in 1981 and did not achieve market success due in part to its $75,000 starting price. As a personal computer, the Xerox Star was rapidly eclipsed by the IBM-PC, the very successful MS-DOS-based personal computer launched in 1981 without a GUI at a price of $1,595.

It is well-known that Steve Jobs and a group of Apple Computer employees made a fortuitous visit to Xerox PARC in December 1979 and received an inside look at the Alto and its GUI. Upon seeing the Alto's user interface, Jobs has been quoted as saying, "It was like a veil being lifted from my eyes. I could see the future of what computing is destined to be."

Much of what Jobs and his team learned at PARC—in exchange for the purchase of one hundred thousand Apple shares by Xerox—was incorporated into the unsuccessful Apple Lisa computer (1982) and later the popular Macintosh (1984). The Apple engineers also implemented features that further advanced the GUI in ways that the PARC researchers had not thought of or were unable to accomplish. Apple Computer was so successful at implementing a GUI-based personal computer that many of the Xerox engineers left PARC and joined Steve Jobs, including Alan Kay and several of his team members.

In response to both the popularity and ease-of-use superiority of the GUI, Microsoft launched Windows in 1985 for the IBM-PC and PC clone markets. The early Windows interface was plagued with performance issues due in part to the fact that it was running as a second layer of programming on top of MS-DOS. With Windows 95, Microsoft developed perhaps the most successful GUI-based personal computer software up to that point.

But by 1988, the GUI had become such an important aspect of personal computing that Apple filed a lawsuit against Microsoft for copyright infringement. In the end, the federal courts ruled against Apple in 1994 saying that "patent-like protection for the idea of the graphical user interface or the idea of the desktop metaphor" was not available. Much of Apple's case revolved around defending as its property something called the "look and feel" of the Mac desktop. While rejecting most of Apple's arguments, the court did grant ownership of the trash can icon, upon which Microsoft began using the recycling bin instead.

When looking back today, it is remarkable how the basic desktop and user-experience design that was developed at Xerox PARC in the 1970s has remained the same ever since. Color and shading have been added to make the icons more photographic and the folders and windows more dimensional. However, the essential user elements, visual indicators, scroll bars, etc. have not changed much.

With the advent of mobile (smartphone and tablet) computing, the GUI began to undergo more significant development. With the original iOS on the iPhone and iPod touch, Apple relied heavily upon so-called skeuomorphic GUI design—that is, icons and images that emulate physical objects in the real world, such as a textured bookcase to display e-books in the iBook app.

Competitors—such as those with Android-based smartphones and tablets—have largely copied Apple's mobile GUI approach. Beginning with iOS 7, however, Apple has moved aggressively away skeuomorphic elements in favor of flattened and less pictorial icons and frames etc.

Multi-touch and gesture-based technology—along with voice user interface (VUI)—represent practical evolutionary steps in the progress of human-computer interaction. Swipe, pinch, and rotate have become just as common for mobile users today as double-click, drag-and-drop, and copy-and-paste were for the desktop generation. The same can be said of the haptic experience—tactile feedback such as vibration or rumbling on a controller—of VR and gaming systems that millions of young people are familiar with all over the world.

It is safe to say that it was the pioneering work of the research group at Xerox PARC that made computing something that everyone can do all the time. They were people with big ideas and big goals. In a 1977 article for *Scientific American*, Alan Kay wrote, "How can communication with computers be enriched to meet the diverse needs of individuals? If the computer is to be truly 'personal,' adult and child users must be able to get it to perform useful activities without resorting the services of an expert. Simple tasks must be simple, and complex ones must be possible."

Appendix 5

Books, E-Books, and the E-Paper Chase

In November 2015, Amazon opened its first retail bookstore in Seattle near the campus of the University of Washington. More than two decades after it pioneered online book sales—and initiated the e-commerce disruption of the retail industry—the company seemed to be taking a step backward with its "brick and mortar" Amazon Books.

However, Amazon launched its store concept with a nod to traditional consumer shopping habits—that is, the ability to "kick the tires." Amazon knows very well that many customers like to browse the shelves in bookstores and fiddle with electronic gadgets like the Kindle, Fire TV, and Echo before they make buying decisions.

Some unique features of the Amazon.com buying experience have been extended to the bookstore. Customer star ratings and reviews are posted near book displays; shoppers are encouraged to use the Amazon app and scan barcodes to check prices. Amazon's bookstore initiative was also possibly motivated by the persistence and strength of the print-book market. Despite the rapid rise of e-books, print books have shown a resurgence. Following a sales decline of fifteen million print books in 2013 to just above five hundred million units, there was an increase to 560 million in 2014 and 570 million in 2015. Meanwhile, the American Booksellers Association (ABA) reported a substantial increase in independent bookstores between 2010 and 2015. The ABA reported 1,712 member stores in 2,227 locations in 2015, up from 1,410 in 1,660 locations in 2010.

The ratio of e-book to print-book sales appeared to have leveled off at around one to three. This relationship supported public perception surveys and learning studies that showed the reading experience and information retention properties of print books are superior to that of e-books. The reasons for the uptick in print sales and the slowing of e-book expansion are complex. Changes in the overall economy, adjustments to bookstore inventory from digital print technologies, and the acclimation of consumers to the differences between the two media platforms have created a dynamic and rapidly shifting landscape.

As many analysts have insisted, it is difficult to make any hard and fast predictions about future trends of either segment of the book market. However, two things are clear: (1) the printed book will undergo little further evolution, and (2) the e-book is headed for rapid and dramatic innovation. Amazon launched the e-book revolution in 2007 with the first Kindle device. Although digital books were previously available in various computer file formats and media types like CD-ROMs for decades, e-books connected with Amazon's Kindle took off in popularity beginning in 2008. The most important technical innovation of the Kindle—and a major factor in its success—was the implementation of the e-paper display.

Distinct from backlit LCD displays on most mobile devices and personal computers, e-paper displays are designed to mimic the appearance of ink on paper. Another important difference is that the energy requirements of e-paper devices are significantly lower than LCD-based systems. Even in later models that offer automatic backlighting for low-light reading conditions, e-paper devices will run for weeks on a single charge while most LCD systems require a recharge in less than twenty-four hours.

The theory behind the Kindle's ink-on-paper emulation was originated in the 1970s at the Xerox Palo Alto Research Center (PARC) in California. Nick Sheridon developed his concepts while working to overcome limitations with the displays of the Xerox Alto, the first desktop computer. The early monitors could only be viewed in darkened office environments because of insufficient brightness and contrast.

Sheridon sought to develop a display that could match the contrast and readability of black ink on white paper. Along with his team of engineers at Xerox, Sheridon developed Gyricon, a substrate with thousands of microscopic plastic beads—each of which were half black and half white—suspended in a thin and transparent silicon sheet. Changes in voltage polarity caused either the white or black side of the beads to rotate up and display images and text without backlighting or special ambient light conditions.

After Xerox canceled the Alto project in the early 1980s, Sheridon took his Gyricon technology in a new direction. By the late 1980s, he was working on methods to manufacture a new digital display system as part of the "paperless office." As Sheridon explained later, "There was a need for a paperlike electronic display—e-paper! It needed to have as many paper properties as possible because ink on paper is the 'perfect display.'"

In 2000, Gyricon LLC was founded as a subsidiary of Xerox to develop commercially viable e-paper products. The start-up opened manufacturing facilities in Ann Arbor, Michigan, and developed several products including e-signage that utilized Wi-Fi networking to remotely update messaging. Unfortunately, Xerox shut down the entity in 2005 due to financial problems.

Among the challenges Gyricon faced were making a truly paperlike material that had sufficient contrast and resolution while keeping manufacturing costs low. Sheridon maintained that e-paper displays would only be viable economically if units were sold for less than $100 so that "nearly everyone could have one."

As Sheridon explained in a 2009 interview, "The holy grail of e-paper will be embodied as a cylindrical tube, about 1 centimeter in diameter and 15 to 20 centimeters long, that a person can comfortably carry in his or her pocket. The tube will contain a tightly rolled sheet of e-paper that can be spooled out of a slit in the tube as a flat sheet, for reading, and stored again at the touch of a button. Information will be downloaded—there will be simple user interface—from an overhead satellite, a cell phone network, or an internal memory chip."

By the 1990s, competitors began entering the e-paper market. E Ink, founded in 1998 by a group of scientists and engineers from MIT's Media Lab including Russ Wilcox, developed a concept like Sheridon's. Instead of using rotating beads with white-and-black hemispheres, E Ink introduced a method of suspending microencapsulated cells filled with both black-and-white particles in a thin transparent film. Electrical charges to the film caused the black or white particles to rise to the top of the microcapsules and create the appearance of a printed page.

E Ink's e-paper technology was initially implemented by Sony in 2004 in the first commercially available e-reader called LIBRIe. In 2006, Motorola integrated an E Ink display in its F3 cellular phone. A year later, Amazon included E Ink's 6-inch display in the first Amazon Kindle which became by far the most popular device of its kind.

Subsequent generations of Kindle devices have integrated E Ink displays with progressively improved contrast, resolution, and energy consumption. By 2011, the third-generation Kindle included touch-screen capability (the original Kindle had an integrated hardware keyboard for input).

The third generation of the Kindle Paperwhite combines backlighting and a touch interface with E Ink Carta technology and a resolution of 300 pixels per inch. Many other e-readers, such as the Barnes & Noble Nook, the Kobo, the Onyx Boox, and the PocketBook, also use E Ink products for their displays.

The quest to replicate, as closely as possible in electronic form, the appearance of ink on paper is logical enough. In the absence of a practical and culturally established form, the new media naturally strives to emulate that which came before it. This process is reminiscent of the evolution of the first printed books. For many decades, print carried over the characteristics of the books that were hand-copied by scribes.

It is well-known that Gutenberg's "mechanized handwriting" invention sought to imitate the best works of the Medieval monks. The *Gutenberg Bible*, for instance, has two columns of print text while everything else about the volume—paper, size, ornamental drop caps, illustrations, gold-leaf accents, binding, etc.—required techniques that preceded the invention of printing. Thus, the initial impact of Gutenberg's system was an increase in the productivity of book duplication and the displacement of scribes; it would take some time for the implications of the new process to work its way through the function, form, and content of books.

More than a half century later—following the spread of Gutenberg's invention to the rest of Europe—the book began to evolve dramatically and take on attributes specific to printing and other changes taking place in society. For example, by the first decade of the 1500s, books were no longer stationary objects to be read in exclusive libraries and reading rooms of the privileged few. As their cost dropped, editions became more plentiful and literacy expanded, books were being read everywhere and by everybody.

By the middle 1500s, both the form and content of books became transformed. To facilitate their newfound portability, the size of books fell from the folio (14.5" × 20") to the octavo dimension (7" × 10.5"). By the beginning of the next century, popular literature—the first European novel is widely recognized as Cervantes's *Don Quixote* of 1605—supplanted verse and classic texts. New forms of print media developed such as chapbooks, broadsheets, and newspapers.

It seems clear that the dominance of LCD displays on computers, mobile, and handheld devices is a factor in the persistent affinity of the public for print books. Much of the technology investment and advancement of the past decade—coming from companies such as Apple—has been committed to computer miniaturization and mobility, not the transition from print to electronic media. While first decade e-readers have made important strides, most e-books are still being read on devices that are visually distant from print books, impeding a more substantial migration to the new media.

Additionally, most current e-paper devices have many unpaper-like characteristics, such as relatively small size, inflexibility, limited bit depth, and the inability to write on them. All current model e-paper Kindles, for example, are limited to 6-inch displays with sixteen gray levels beneath a heavy and fragile layer of glass and no support for handwriting.

A new generation of e-paper systems has been developed that overcome many of these limitations. In 2014, Sony released its Digital Paper System (DPT-S1) that is a letter-size e-reader and e-notebook (for $1,100 at launch). The DPT-S1 is based on E Ink's Mobius display, a 13.3" thin film transistor (TFT) platform that is flexible and can accept handwriting from a stylus. Since it does not have any glass, the Sony device weighs 12.6 oz or about half of a similar LCD-based tablet. With the addition of stylus-based handwriting capability, the device functions like an electronic notepad, and meanwhile, notes can be written in the margins of e-books and other electronic documents.

These advancements and others show that e-paper is positioned for a renewed surge into things that have yet to be conceived. Once a flat surface can be curved or even folded and then made to transform itself into any image, including a color image, at any time and at very low cost and very low energy consumption, then many things like e-wallpaper, e-wrapping paper, e-milk cartons, and e-price tags will be developed and the possibilities become nearly limitless.

Appendix 6

The Pioneers of Color Printing

Color is a perception. It is the response of the human visual system to light reflected from objects in the world around us. We learn as children to associate these color perceptions with names: the red of an apple, the green of the leaves, or the blue of the sky. More scientifically, color is the way our eyes, optic nerve, and brain receive and process different wavelengths of the visible spectrum of electromagnetic radiation.

It took hundreds of years of thought and experiment—beginning with Isaac Newton's 1672 idea that white light is the source of color sensation—to arrive at the modern understanding of color and the way we perceive it. In 1802, the visible spectrum of electromagnetic energy was defined by Thomas Young when he measured various wavelengths of light and established their relationship to color—that is, red is about 650 nm, green is about 510 nm, and blue is about 475 nm.

Later, Young and Hermann von Helmholtz developed the theory of trichromatic color vision. They surmised that the human eye has three types of photoreceptors, each with sensitivity to a corresponding range of light waves. In the 1950s, it was proved—with advanced measuring equipment—that the three kinds of visual receptors (cones) have the capacity to sense many combinations of light wavelengths and see them as all the colors of the rainbow.

Knowledge of the properties of light and color was a major achievement of the scientific revolution of the eighteenth and nineteenth century that—alongside the discovery of graphical perspective and other mathematical linear projections—transformed the visual arts. Artists and craftsmen that exclusively relied on their sensibility, talent, and visual experience were able to integrate the principles of science into their works, bringing a degree of realism and accuracy that was previously impossible.

While the history of two-dimensional color representation is most often associated with fine art painting—especially the fresco, oil, and tempera works of the Renaissance masters—the lesser-known origins of color printing took a parallel development in time. Starting with the birth of mechanical metal type in Germany during the High Renaissance, a quest was begun to conquer the challenge of practical and high-quality color printing.

Relief color: Fust and Schöffer (1457)

There is evidence that Johannes Gutenberg experimented with color during the printing of his famous forty-two-line Bible. For the most part, however, traditional hand-painted ornamental color lettering was used by Gutenberg alongside the black letter printing type he invented around 1440. Shortly thereafter, relief printing of color type and other ornamental figures was performed remarkably well by Gutenberg's former collaborators on the printing of the Bible, Johann Fust and Peter Schöffer. In 1457, Fust and Schöffer printed the Mainz Psalter with three colors—black, red, and blue—all at one time. Their ingenious technique of compound printing involved interlocking metal-type characters that were inked separately and reassembled for a single impression on the printing press. Although it returned extraordinary results, the process was very time-consuming and expensive.

Intaglio color: Teyler (1680)

For most of the next two centuries, limited color printing was attempted while hand-colored pages remained the preferred method of pictorial representation. As various printing techniques spread across Europe, new approaches to color reproduction were tested. Some color illustrations were made using woodblocks. By the mid-fifteenth century, intaglio engraving emerged as the standard method for printing images in black and white. Around 1680, a mathematician and engineer from Nijmegen, Holland, named Johan Teyler developed a means of dabbing different colored inks into the wells of intaglio plates—originally intended for black-only printing—to make a full-color impression all at one time. Like Fust and Schöffer's work, Teyler's color results were artistically beautiful but could not be developed into a viable commercial process.

Three-color mezzotint: Le Blon (1720)

Following the publication in 1704 of Isaac Newton's discoveries regarding the physics of light and color—especially the idea that all colors are made of different combinations of red, blue, and yellow (this was later proven to be imprecise for both light waves and pigments)—a few printers began working with techniques in three-color mezzotint printing. By this time, mezzotint copperplates were the favored image reproduction method because they rendered tones more easily than engraving. In 1711, the Frankfurt-born painter Jacob Christoph Le Blon, basing himself directly upon Newton's theory, mastered trichromatic mezzotint printing in his Amsterdam studio. Le Blon first tried and failed to commercialize his invention in Amsterdam, the Hague, and Paris. He relocated in London in 1720, successfully obtained a royal patent from George I for his process, and opened a business selling printed color copies of oil paintings. While his technical accomplishment was a significant step forward, Le Blon's business lasted for three years before bankruptcy forced him back to Paris. Creating an appropriate balance of intensity between the primary color plates and maintaining tight registration between the three press impressions upon the paper was an exceedingly difficult and costly trial-and-error process.

Four-color mezzotint: L'Admiral and Gautier (1736)

It is documented that J. C. Le Blon also invented four-color mezzotint printing by adding black to the red, yellow, and blue plates on a few of his prints. However, the perfection RYBK (K is for the key color, black) model was made by others, especially following Le Blon's death in 1741. Among those who contributed were the Dutch engraver and printer Jan L'Admiral and the French painter Jacques Gautier, who had both been assistants to Le Blon. L'Admiral and Gautier initially produced color plates of human anatomy for medical research publications in Paris in the 1730s and 1740s. In France, Gautier proved to be something of a charlatan and attempted to take full credit for the invention of four-color printing. He started a periodical in 1752 called *Observations* on natural history, on physics, and painting in which popular sensationalism appears to have been his primary objective. Gautier fabricated a full-color image of a "siren" with the body of a seahorse and a hideous head of a human and reproduced some images that bordered on pornography. Nonetheless, Gautier's journal was among the first financially successful uses of color printing and it was published quarterly for five years.

Chromolithography: Engelmann (1837)

The invention of lithography by the Bavarian Alois Senefelder in 1796 brought a fundamentally new approach to printing. While relief letterpress and intaglio mezzotint were mechanical printing processes, lithography relied upon the chemical antipathy of oil and water to transfer the image onto paper. The new method enabled artists to draw on the surface of limestone instead of the much more difficult etching or engraving of metal plates. In 1818, Senefelder experimented with lithographic color reproduction and pointed the way forward for others. For the next two decades, lithographers from Germany, France, and England made strides with color for the most part printing decorative ornamentations or title pages of relief printed books. The chromolithography during this time was also a return to a multicolor approach of the seventeenth century as opposed to the later and more advanced three- or four-color mezzotint separation process. Chromolithography came of age with the work of the French-German Godefroy Engelmann of Mulhouse. After becoming a pioneer and master in monochrome lithography, Engelmann made significant progress with four-color lithographs in 1837. He moved to Paris a year later and obtained a patent for his process. His works proved that chromolithography could effectively render lifelike prints of landscapes, flower, and fruit arrangements and the most difficult human forms.

The four-color chromolithography of the mid-nineteenth century finally brought color printing to an economically viable balance of quality, time, and cost. However, full color printing was still largely a special process that was performed separately from letterpress black-only text print. With the rapid industrial expansion of book and newspaper publishing, color work remained essentially a very slow, craft-based process that required highly skilled artisans.

The production of relief, intaglio, and lithographic "prints" and "plates" remained the convention for color work during the balance of the 1800s. These products were most often sold as single items—sometimes for as little as a penny each—or bound into books as illustrations. It would require the development and maturity of three major advancements in the graphic arts to integrate printed color together with black text: color photography by Thomas Sutton (1861) and halftone reproduction by Frederic Ives (1881) and the CMYK ink model (1906).

Initially, some black-and-white halftones were enhanced with synthetically applied color. By the 1920s, improvements in mechanical color separation techniques and the growth of magazine publishing made it possible for some titles to afford full-color pictures and black text to be printed together on sheetfed letterpress systems. Some of these publications, such as *National Geographic* magazine, continued with letterpress color all the way up to the late 1970s.

By the late 1950s, offset lithography and electronic color separations had begun their rise as the foremost method of reproducing high quality, inexpensive printed color images. Although the personal computer (IBM, 1981), digital camera (Fuji, 1987), and digital printing (Indigo, 1993) have brought color reproduction to new levels of high quality and low cost—especially for small quantities— the breakthroughs of seventy years ago remain by far the dominant methods of color reproduction today.

Appendix 7

How Index Cards Launched
the Information Age

On October 1, 2015, the final order of library catalog cards was printed by the Online Computer Library Center (OCLC) in Dublin, Ohio. The next day, *The Columbus Dispatch* wrote, "Shortly before 3:00 p.m. Thursday, an era ended. About a dozen people gathered in a basement workroom to watch as a machine printed the final sheets of library catalog cards to be made…" The demise of the printed library card, an indispensable indexing tool for more than a century, was inevitable in the age of electronic information and the Internet. It is safe to say that nearly all print with purely informational content—as opposed to items fulfilling a promotional or a packaging function—is surely to be replaced by online alternatives.

Founded in 1967, the OCLC is a global cooperative with sixteen-thousand-member libraries. Although it no longer prints library cards, the OCLC continues to fulfill its mission by providing shared library resources, such as catalog metadata and WorldCat.org, an international online database of library collections. Speaking about the end of the card catalog era, Skip Prichard the CEO of the OCLC said, "The vast majority of libraries discontinued their use of the printed library catalog card many years ago…. But it is worth noting that these cards served libraries and their patrons well for generations, and they provided an important step in the continuing evolution of libraries and information science."

The library catalog card is one form of the popular 3 × 5 index card that served as a filing system for a multitude of purposes for over two hundred years. While some of us have been around long enough to have used or maybe even still use them—for addresses and phone numbers, recipes, flash cards, or research paper outlines—we may not be aware of the relationship that index cards have to modern information science. The original purpose of the index card and its subsequent development represented the early stages of information theory and practice. Additionally, as becomes clear below, without the index card as the first functional system for organizing complex categories, subcategories, and cross-references, studies in the natural sciences would have never gotten off the ground.

The index card became the indispensable tool for both organizing and comprehending the expansion of human knowledge at every level. Along with several important intermediary steps, the ideas that began with index cards eventually led to relational databases, document management systems, hyperlinks, and the World Wide Web.

The Swedish naturalist and physician Carl Linnaeus (1707–1778) is recognized as the creator of the index card. Linnaeus used the cards to develop his system of organizing and naming the species of all living things. Linnaean taxonomy is based on a hierarchy (kingdom, phylum, class, order, family, genus, species) and binomial species naming (*Homo erectus*, tyrannosaurus rex, etc.). He published the first edition of his universal conventions in a small pamphlet called *The System of Nature* in 1735.

Beginning in his early twenties, Linnaeus was interested in producing a series of books on all known species of plants and animals. At that time, there were so many new species being discovered that Linnaeus knew as soon as a book was printed, a large amount of new information would already be available. He wanted to quickly and accurately revise his publications to incorporate the new findings in subsequent editions. As time went on, Linnaeus developed different functional methods of sorting through and organizing enormous amounts of information connected with his growing collection of plant, animal, and shell specimens (eventually, it rose to forty thousand samples). His biggest problem was creating a process that was both structured enough to facilitate retrieval of previously collected information and flexible enough to allow rearrangement and addition of new information.

Working with paper notations in the eighteenth century, he needed a system that would allow the flow of names, references, descriptions, and drawings into and out of a fixed sequence for the purposes of comparison and rearrangement. This "packing" and "unpacking" of information was a continuous process that enabled Linnaeus's research to keep up with the changes in what was known about living species. At first, Linnaeus used notebooks. This linear method—despite his best efforts to leave pages open for updates and new information—proved to be unworkable and wasteful. As estimates of how much room to allow often proved incorrect, Linnaeus was forced to squeeze new details into ever shrinking available space, or he ended up with unutilized blank pages.

After thirty years of working with notebooks, Linnaeus began to experiment with a filing system of information recorded on separate sheets of paper. This was later converted to small sheets of thick paper that could be quickly handled, shuffled through, and laid out on a table in two-dimensions like a deck of playing cards. This is how the index card was born.

Linnaeus's index card system was able to represent the variation of living organisms by showing multiple affinities in a maplike fashion. To accommodate the ever-expanding knowledge of new species—today the database of taxonomy contains 8.7 million items—Linnaeus created a breakthrough method for managing complex information.

While index cards continued to be used in Europe, an important step forward in information management was made in the US by Melvil Dewey (1851–1931), the creator of the well-known Dewey Decimal System (or Dewey Decimal Classification, DDC). Used by libraries for the cataloging of books since 1876, the DDC was based on index cards and introduced the concepts of "relative location" and "relative index" to bibliography. It also enabled libraries to add books to their collection based on subject categories and an infinite number of decimal expressions known as "call numbers."

Before the DDC, libraries attempted to assign books to a permanent physical location based on their order of acquisition. This linear approach proved unworkable, especially as library collections grew rapidly in the latter part of the nineteenth century. With industrialization, libraries were overflowing with paper: letters, reports, memos, pamphlets, operation manuals, schedules, as well as books were flooding in, and the methods of cataloging and storing these collections needed to find a means of keeping up.

In the 1870s, while working at Amherst College Library, Melvil Dewey became involved with libraries across the country. He was a founding member of the American Library Association and became editor of *The Library Journal*, a trade publication that still exists today. In 1878, Dewey published the first edition of *A Classification and Subject Index for Cataloguing and Arranging the Books and Pamphlets of a Library* that elaborated on the use of the library card catalog index.

Like many others of his generation, Melvil Dewey was committed to scientific management, standardization and the democratic ideal. By the end of the nineteenth century, the Dewey classification system and his 3 × 5 card catalog were being used in nearly every school and public library in the US. The basic concept was that any member of society could walk into a library anywhere in the country, go to the card catalog, and be able to locate the information they were looking for.

In 1876, Dewey created a company called Library Bureau and began providing card catalog supplies, cabinets, and equipment to libraries across the country. Following the enormous success of this business, Dewey expanded the Library Bureau's information management services to government agencies and large corporations at the turn of the twentieth century. In 1896, Dewey formed a partnership with Herman Hollerith and the Tabulating Machine Company (TMC) to provide the punch cards used for the electromechanical counting system of the US government census operations. Dewey's relationship with Hollerith is significant as TMC would be renamed International Business Machines (IBM) in 1924 and become an important force in the information age and creator of the first relational database.

While Dewey's classification system became the standard in US libraries, others were working on bibliographic cataloging ideas, especially in Europe. In 1895, the Belgians Paul Otlet (1868–1944) and Henri La Fontaine founded the International Institute of Bibliography (IIB) and began working on something they called the Universal Bibliographic Repertory (UBR), an enormous catalog based on index cards. Funded by the Belgian government, the UBR involved the collection of books, articles, photographs, and other documents in order to create a one-of-a-kind international index.

As described by Otlet, the ambition of the UBR was to build "an inventory of all that has been written at all times, in all languages, and on all subjects." Although they used the DDC as a starting point, Otlet and La Fontaine found limitations in Dewey's classification system while working on the UBR. Some of the issues were related to Dewey's American perspective; the DDC lacked some categories needed for information related to other languages and regions of the world.

More fundamentally, however, Otlet and La Fontaine made an important conceptual breakthrough over Dewey's approach. They conceived of a complex multidimensional indexing system that would allow for more deeply defined subject categories and cross-referencing of related topics. Their critique was based on Otlet's pioneering idea that the content of bibliographic collections needed to be separated from their form and that a "universal" classification system needed to be created that included new media and information sources (magazines, photographs, scientific papers, audio recordings, etc.) and moved away from the exclusive focus on the location of books on library shelves.

After Otlet and La Fontaine received permission from Dewey to modify the DDC, they set about creating the Universal Decimal Classification (UDC). The UDC extended Dewey's cataloging expressions to include symbols (equal sign, plus sign, colon, quotation marks, and parenthesis) for the purpose of establishing "links" between multiple topics. This was a very significant breakthrough that reflected the enormous growth of information taking place at the end of the nineteenth century.

By 1900, the UBR had more than three million entries on index cards and was supported by more than three hundred IIB members from dozens of countries. The project was so successful that Otlet began working on a plan to copy the UBR and distribute it to major cities around the world. However, with no effective method for reproducing the index cards, other than typing them out by hand, this project ran up against the technical limitations of the time. In 1910, Otlet and La Fontaine shifted their attention to the establishment of the Mundaneum in Mons, Belgium. Again, with government support, the aim of this institution was to bring together all the world's knowledge in a single UDC index. They created the gigantic repository as a service where anyone in the world could submit an inquiry on any topic for a fee. This analog search service would provide information back to the requester in the form of index cards copied from the Mundaneum's bibliographic catalog. By 1924, the Mundaneum contained eighteen million index cards housed in fifteen thousand catalog drawers. Plagued by financial difficulties and a reduction of support from the Belgian government during the Depression and lead up to World War II, Paul Otlet realized that further management of the card catalog had become impractical. He began to consider more advanced technologies—such as photomechanical recording systems and even ideas for electronic information sharing—to fulfill his vision.

Although the Mundaneum was sacked by the Nazi's in 1940 and most of the index cards destroyed, the ideas of Paul Otlet anticipated the technologies of the information age that were put into practice after the war. The pioneering work of others—such as Emanuel Goldberg, Vannevar Bush, Douglas Engelbart, and Ted Nelson—would lead to the creation of the Internet, World Wide Web, and search engines in the second half of the twentieth century.

Photos and Graphics

Pages 44-45: Framework for technological disruptive continuity in the age of print. Original graphic by Kevin Reed Donley.

Page 66: From an engraved portrait of Johannes Gutenberg made after his death in 1468 by an anonymous and unknown artist. (https://commons.wikimedia.org/wiki/File:Gutenberg.jpg), "Gutenberg", Cropped, converted to grayscale and mosaic filter applied by Kevin Reed Donley, https://creativecommons.org/publicdomain/zero/1.0/legalcode

Page 70: From an engraved portrait of Nicolas Jenson by an anonymous and unknown artist. (https://commons.wikimedia.org/wiki/File:Portrait_of_Nicholas_Jenson.gif), "Portrait of Nicholas Jenson", Cropped, converted to grayscale and mosaic filter applied by Kevin Reed Donley, https://creativecommons.org/publicdomain/zero/1.0/legalcode

Page 76: From an engraving of Aldus Manutius entitled, *Life of Aldus Pio Manutius* in 1759 by anonymous and unknown artist. (https://commons.wikimedia.org/wiki/File:Vita_di_Aldo_Pio_Manuzio_(cropped).jpg), "Vita di Aldo Pio Manuzio (cropped)", Cropped, converted to grayscale and mosaic filter applied by Kevin Reed Donley, https://creativecommons.org/publicdomain/zero/1.0/legalcode

Page 82: From an oil on canvas portrait of Benjamin Franklin by Joseph-Siffred Duplessis in 1783.
After Joseph-Siffred Duplessis artist QS:P170,Q4233718,P1877, Q2286906 (https://commons.wikimedia.org/wiki/File:Benjamin_ Franklin_by_Joseph_Siffrein_Duplessis.jpg), "Benjamin Franklin by Joseph-Siffred Duplessis", Cropped, converted to grayscale and mosaic filter applied by Kevin Reed Donley, https://creativecommons.org/publicdomain/zero/1.0/legalcode

Page 86: From an oil on canvas portrait of William Blake by Thomas Phillips in 1807.
Thomas Phillips artist QS:P170,Q422726 (https://commons.wikimedia.org/wiki/File:William_Blake_by_Thomas_Phillips.jpg), "William Blake by Thomas Phillips", Cropped, converted to grayscale and mosaic filter applied by Kevin Reed Donley, https://creativecommons.org/publicdomain/zero/1.0/legalcode

Page 92: From a portrait of Nicolas-Louis Robert painted by his sister circa 1780.
Chienlit (https://commons.wikimedia.org/wiki/File:Louis-Nicolas_ Robert_30_painted_by_his_sister.jpg), "Louis-Nicolas Robert 30 painted by his sister", Cropped, converted to grayscale and mosaic filter applied by Kevin Reed Donley, https://creativecommons.org/ publicdomain/zero/1.0/legalcode

Page 100: From a lithograph portrait of Alois Senefelder by Nicolas Henri Jacob and printed by Gottfried Engelmann in 1819.
Nicolas Henri Jacob (https://commons.wikimedia.org/wiki/ File:Alois_Senefelder_Portrait.jpg), "Alois Senefelder Portrait", Cropped, converted to grayscale and mosaic filter applied by Kevin Reed Donley, https://creativecommons.org/publicdomain/zero/1.0/ legalcode

Page 104: From an oil painting of Charles Stanhope, 3rd Earl Stanhope by John Opie.

John Opie artist QS:P170,Q2481088 (https://commons.wikimedia.
org/wiki/File:Charles_Stanhope,_3rd_Earl_Stanhope_by_John_
Opie.jpg), "Charles Stanhope, 3rd Earl Stanhope by John Opie",
Cropped, converted to grayscale and mosaic filter applied by Kevin
Reed Donley, https://creativecommons.org/publicdomain/zero/1.0/
legalcode

Page 112: From a lithograph of Friedrich Koenig in 1900 by an
anonymous and unknown artist. (https://commons.wikimedia.org/
wiki/File:Friedrich_koenig.jpg), "Friedrich Koenig", Cropped, converted to grayscale and mosaic filter applied by Kevin Reed Donley,
https://creativecommons.org/publicdomain/zero/1.0/legalcode

Page 116: From a photographic portrait of George Baxter supplied
by Mr. Frederick Harrild, taken from the original in the possession of
the family, by an anonymous and unknown artist. (https://commons.
wikimedia.org/wiki/File:George_Baxter_from_family_photograph.
jpg), "George Baxter from family photograph", Cropped, converted
to grayscale and mosaic filter applied by Kevin Reed Donley,
https://creativecommons.org/publicdomain/zero/1.0/legalcode

Page 124: From a photographic portrait of Richard March Hoe by
Mathew Benjamin Brady and held by the Library of Congress.
Mathew Benjamin Brady creator QS:P170,Q187850 (https://
commons.wikimedia.org/wiki/File:Richard_March_Hoe_-_Brady-
Handy.jpg), "Richard March Hoe–Brady-Handy", Cropped, converted to grayscale and mosaic filter applied by Kevin Reed Donley,
https://creativecommons.org/publicdomain/zero/1.0/legalcode

Page 130: From a photographic portrait of William Bullock found
in a family album by an anonymous and unknown artist. (https://
commons.wikimedia.org/wiki/File:William_bullock_inventor_por-
trait.jpg), "William bullock inventor portrait", Cropped, converted
to grayscale and mosaic filter applied by Kevin Reed Donley,
https://creativecommons.org/publicdomain/zero/1.0/legalcode

Page 134: From a photographic portrait of Christopher Latham Sholes contained in *Leading American Inventors* by George Iles and published in 1912.
Iles, George (https://commons.wikimedia.org/wiki/File:Sholes.jpg), "Sholes", Cropped, converted to grayscale and mosaic filter applied by Kevin Reed Donley, https://creativecommons.org/publicdomain/zero/1.0/legalcode

Page 140: From a photographic self-portrait of Frederic Eugene Ives taken c. 1900-1905.
Rijksmuseum (https://commons.wikimedia.org/wiki/File:Portret_van_Frederic_Eugene_Ives,_RP-F-2001-7-509-107.jpg), "Portret van Frederic Eugene Ives, RP-F-2001-7-509-107", Cropped, converted to grayscale and mosaic filter applied by Kevin Reed Donley, https://creativecommons.org/publicdomain/zero/1.0/legalcode

Page 144: From a printed advertisement for the Edison-Dick Mimeograph in the magazine *Collier's: The National Weekly* dated June 14, 1919, by an anonymous and unknown artist.
(https://commons.wikimedia.org/wiki/File:Grist!—Edison-Dick_Mimeograph.jpg), "Grist!—Edison-Dick Mimeograph", Cropped and converted to grayscale by Kevin Reed Donley, https://creativecommons.org/publicdomain/zero/1.0/legalcode

Page 150: From an etched portrait of Linn Boyd Benton by R.H. Sommer as published by *The Inland Printer* in 1922.
R. H. Sommer (https://commons.wikimedia.org/wiki/File:Linn_Boyd_Benton_by_R._H._Sommer.jpg), "Linn Boyd Benton by R. H. Sommer", Cropped, converted to grayscale and mosaic filter applied by Kevin Reed Donley,
https://creativecommons.org/publicdomain/zero/1.0/legalcode

Page 158: From an engraved portrait of Ottmar Mergenthaler by E.G. Williams and Bro. in 1894.
From JHU Sheridan Libraries/Gado/Getty Images (https://commons.wikimedia.org/wiki/File:Ottmar_Mergenthaler,_1894.jpg),

"Ottmar Mergenthaler, 1894", Cropped, converted to grayscale and mosaic filter applied by Kevin Reed Donley, https://creativecommons.org/publicdomain/zero/1.0/legalcode

Page 162: From a photographic portrait of William Morris by Frederick Hollyer c. 1887.
Frederick Hollyer creator QS:P170,Q3496914 (https://commons.wikimedia.org/wiki/File:William_Morris_age_53.jpg), "William Morris age 53", Cropped, converted to grayscale and mosaic filter applied by Kevin Reed Donley, https://creativecommons.org/publicdomain/zero/1.0/legalcode

Page 170: From a photographic portrait of Karel Klíč by an anonymous and unknown artist. (https://commons.wikimedia.org/wiki/File:Karel_Václav_Klíč_(1841-1926).jpg), "Karel Václav Klíč (1841-1926)", Cropped, converted to grayscale and mosaic filter applied by Kevin Reed Donley, https://creativecommons.org/publicdomain/zero/1.0/legalcode

Page 176: From a photographic portrait of Frederick Goudy by Arnold Genthe c. 1924.
Genthe, Arnold (https://commons.wikimedia.org/wiki/File:Portrait_photograph_of_Frederick_W._Goudy_LOC_agc.7a10228.jpg), "Portrait photograph of Frederick W. Goudy LOC agc.7a10228", Cropped, converted to grayscale and mosaic filter applied by Kevin Reed Donley, https://creativecommons.org/publicdomain/zero/1.0/legalcode

Page 182: From a portrait of Ira Rubel published by Penrose's Pictorial Annual, The Process Yearbook: A Review of the Graphic Arts, Vol. XIV, 1908-09 by an anonymous and unknown artist.
Unknown (https://commons.wikimedia.org/wiki/File:Ira_Washington_Rubel.jpg), "Ira Washington Rubel", Cropped, converted to grayscale and mosaic filter applied by Kevin Reed Donley, https://creativecommons.org/publicdomain/zero/1.0/legalcode

Page 188: From a heliogravure print of a drawing of Stanley Morison by William Rothenstein in 1923.
After William Rothenstein (1872-1945) (https://commons.wikimedia.org/wiki/File:William_Morison_by_Rothenstein.png), "William Morison by Rothenstein", Cropped, converted to grayscale and mosaic filter applied by Kevin Reed Donley,
https://creativecommons.org/publicdomain/zero/1.0/legalcode

Page 194: From a photograph of 23 prominent scientists and engineers, including Chester Carlson, in the National Inventors Council taken on January 1, 1966 by an anonymous and unknown artist.
National Institute of Standards and Technology (https://commons.wikimedia.org/wiki/File:The_National_Inventors_Council,_1966.jpg), "The National Inventors Council, 1966", Cropped, converted to grayscale and mosaic filter applied by Kevin Reed Donley,
https://creativecommons.org/publicdomain/zero/1.0/legalcode

Page 198: From a painted self-portrait of John Crosfield at age 15 in 1931. (https://commons.wikimedia.org/wiki/File:Self_Portrait_John_Crosfield_1931.jpg), "Self Portrait John Crosfield 1931", Cropped, converted to grayscale and mosaic filter applied by Kevin Reed Donley,
https://creativecommons.org/licenses/by-sa/3.0/legalcode

Page 206: From a photograph of Rudolf Hell demonstrating his weather chart recorder at a technical conference in Kiel in 1968 taken by Friedrich Magnussen.
Magnussen, Friedrich (1914-1987) (https://commons.wikimedia.org/wiki/File:Rudolf_Hell_führt_seinen_Wetterkartenschreiber_vor_(Kiel_44.592).jpg), "Rudolf Hell führt seinen Wetterkartenschreiber vor (Kiel 44.592)", Cropped, converted to grayscale and mosaic filter applied by Kevin Reed Donley,
https://creativecommons.org/licenses/by-sa/3.0/de/legalcode

Page 214: From a photograph by Jeremy Norman of *The Wonderful World of Insects*, the first book to be phototypeset with Louis Moyroud's Lumitype machine in 1953.
Jeremy Norman Collection of Images (https://historyofinformation.com/images/Photon_wonderful_world_cover_2021-02-04T08:01:21-08:00_big.jpg), "The Wonderful World of Insects", Cropped, converted to grayscale and mosaic filter applied by Kevin Reed Donley,
https://creativecommons.org/licenses/by-sa/4.0/legalcode

Page 220: From a specimen of the typeface Helvetica, originally designed by Eduard Hoffmann in 1957.
Luisao10   (https://commons.wikimedia.org/wiki/File:Helvetica_tipografia.svg), "Helvetica tipografia", Cropped and converted to grayscale by Kevin Reed Donley, https://creativecommons.org/licenses/by-sa/3.0/legalcode

Page 226: From a specimen of the typeface OCR-B, originally designed by Adrian Frutiger in 1968.
ZoeB   (https://commons.wikimedia.org/wiki/File:Typeface_specimen_OCR_B.svg), Cropped and converted to grayscale by Kevin Reed Donley,
https://creativecommons.org/licenses/by-sa/4.0/legalcode

Page 234: From a photograph of Hermann Zapf signing one of the test glass panels he designed for the Lawson center at RIT during a ceremony on May 11, 2007.
The original uploader was Lovibond at English Wikipedia. (https://commons.wikimedia.org/wiki/File:Hermann_Zapf_signing.jpg), "Hermann Zapf signing", Cropped, converted to grayscale and mosaic filter applied by Kevin Reed Donley, https://creativecommons.org/licenses/by-sa/2.5/legalcode

Page 242: From a photographic portrait of Marshall McLuhan taken by Bernard Gotfryd in 1967. Gotfryd, Bernard, photographer (https://commons.wikimedia.org/wiki/File:Marshall_McLuhan_

with_and_on_television.jpg), "Marshall McLuhan with and on television", Cropped, converted to grayscale and mosaic filter applied by Kevin Reed Donley,
https://creativecommons.org/publicdomain/zero/1.0/legalcode

Page 246: From a typeface specimen of Avant Garde Gothic, originally designed by Herb Lubalin in 1970.
Inferno986return (https://commons.wikimedia.org/wiki/File:ITC_Avant_Garde_Gothic_Sample.svg), Cropped and converted to grayscale by Kevin Reed Donley,
https://creativecommons.org/licenses/by-sa/4.0/legalcode

Page 252: From image 2 (page 3) of the patent application by Robert Howard and others for the dot matrix printer filed in 1970.
US Govt work (https://patents.google.com/patent/US3703949), "From High-Speed Printer US Patent Application", Cropped and converted to grayscale by Kevin Reed Donley, https://creativecommons.org/publicdomain/zero/1.0/legalcode

Page 262: From a family photograph of Efraim "Efi" Arazi as published by Studio Yoram Aschheim by an anonymous and unknown artist.
Studio Yoram Aschheim (https://commons.wikimedia.org/wiki/File:יזרא_יפא.jpg), "יזרא יפא", Cropped, converted to grayscale and mosaic filter applied by Kevin Reed Donley, https://creativecommons.org/licenses/by-sa/3.0/legalcode

Page 268: From a portrait of Steve Jobs and the Macintosh computer by Bernard Gotfryd taken in January 1984.
Photo: Bernard Gotfryd—Edited from tif by Cart (https://commons.wikimedia.org/wiki/File:Steve_Jobs_and_Macintosh_computer,_January_1984,_by_Bernard_Gotfryd_-_edited.jpg), "Steve Jobs and Macintosh computer, January 1984, by Bernard Gotfryd —edited", Cropped, converted to grayscale and mosaic filter applied by Kevin Reed Donley,
https://creativecommons.org/publicdomain/zero/1.0/legalcode

Adams, J. Michael and Penny Ann Dolin, *Printing Technology*, Fifth
    Edition, Albany, NY, Delmar, Thompson Learning, 2002.
Ayala, Francisco J., *Darwin's Gift to Science and Religion*, Washington,
    DC, Joseph Henry Press, 2007.
Baker, Elizabeth Faulkner, *Displacement of Men by Machines: Effects
    of Technological Change in Commercial Printing*, New York, NY,
    Arno Press, 1977.
Barolini, Helen, *Aldus and His Dream Book*, New York, NY, Italica
    Press, 1992.
Basalla, George, *The Evolution of Technology*, New York, NY,
    Cambridge University Press, 1988.
Basbanes, Nicholas A., *On Paper: The Everything of Its Two-Thousand-
    Year History By a Self-Confessed Bibliophiliac*, New York, NY,
    Alfred A. Knopf, 2013.
Blair, Raymond and Charles Shapiro, *The Lithographers Manual*,
    Pittsburgh, PA, The Graphic Arts Technical Foundation, Inc.,
    1980.
Bolter, Jay David, *Writing Space: Computers, Hypertext, and the
    Remediation of Print*, Mahwah, NJ, Lawrence Erlbaum
    Associates, Publisher, 2001.
Brands, H. W., *The First American: The Life and Times of Benjamin
    Franklin*, New York, NY, Anchor Books, 2000.
Briggs, Asa and Peter Burke, *A Social History of the Media from
    Gutenberg to the Internet*, Malden, MA, Polity Press, 2009.

Bringhurst, Robert, *The Elements of Typographic Style*, Point Roberts, WA, Hartley and Marks Publishers, Inc., 1997.

Bruckner, D. J. R., *Frederick Goudy*, New York, NY, Harry N. Abrams, Inc., 1990.

Bryans, Dennis, *A Seed of Consequence: Ph. D. Thesis*, Melbourne, Australia, Swinburne University of Technology, 2000.

Buckland, Michael, *Emanuel Goldberg and His Knowledge Machine: Information, Invention, and Political Forces*, Westport, Connecticut, Libraries Unlimited, 2006.

Burch, R. M., *Colour Printing and Colour Printers*, Edinburgh, Paul Harris Publishing, 1983.

Carter, Sabastian, *Twentieth Century Type Designers*, New York, NY, W. W. Norton and Company, 1995.

Christensen, Edward H., *Fifty Golden Years of Craftsmanship: 1919-1969*, Ephrata, Pennsylvania, The Science Press, Inc., 1969.

Cost, Frank, *The New Medium of Print*, Rochester, NY, RIT Cary Graphic Arts Press, 2005.

Cost, Patricia A., *The Bentons*, Rochester, NY, RIT Cary Graphic Arts Press, 2011.

Cringely, Robert X., *Accidental Empires: How the Boys of Silicon Valley Make Their Millions, Battle Foreign Competition, and Still Can't Get a Date*, Reading, MA, Addison-Wesley Publishing Company, Inc., 1992.

Crosfield C.B.E., D. Sc., M.A., John F., *Recollections of Crosfield Electronics 1947 to 1975*, Peterborough, England, Fisherprint Ltd., 1991.

Derry, T. K. and Trevor I. Williams, *A Short History of Technology: From the Earliest Times to A.D. 1900*, New York, NY, Oxford University Press, 1961.

Distad, Merrill, *George Baxter: Master Colour Printer*, Edmonton, Alberta, University of Alberta Libraries, 2016.

Dobras, Wolfgang, *Gutenberg: Man of the Millennium*, Mainz, Germany, City of Mainz, 2000.

Durant, Will, *The Story of Civilization: The Reformation (Volume VI)*, New York, NY, Simon and Schuster, 1957.

Eckersley, Richard, *The Glossary of Typesetting Terms*, Chicago, IL, The University of Chicago Press, 1994.

Eisenstein, Elizabeth L., *The Printing Press as an Agent of Change*, New York, NY, Cambridge University Press, 1979.

Eisenstein, Elizabeth L., *The Printing Revolution in Early Modern Europe*, New York, NY, Cambridge University Press, 1993.

Ellis, Charles D., *Joe Wilson and the Creation of Xerox*, Hoboken, NJ, John Wiley & Sons, Inc, 2006.

Encyclopaedia Britannica Company, *Graphic Arts*, Garden City, New York, Garden City Publishing Company, 1936.

Erdman, David V., Editor, *The Selected Poetry of William Blake*, New York, NY, Signet Classics, 1976.

Febvre, Lucian and Henri-Jean Martin, *The Coming of the Book*, New York, NY, Verso, 1976.

Finley, Charles, *Printing Paper and Ink*, Albany, NY, Delmar Publishers, 1997.

Foulke, Arthur T., *Mr. Typewriter*, Boston, MA, The Christopher Publishing House, 1961.

Friedl, Friedrich, *The Univers by Adrian Frutiger*, Frankfurt am Main, Verlag Form Gmbh, 1998.

Frutiger, Adrian, *Signs and Symbols: Their Design and Meaning*, New York, NY, Watson-Guptill Publications, 1997.

Gascoigne, Bamber, *Milestones in Colour Printing 1457–1859*, New York, NY, Cambridge University Press, 1997.

Gavelin, Gunner, *Fourdrinier Papermaking*, New York, NY, Lockwood Trade Journal Co., Inc., 1963.

Gill, Eric, *An Essay on Typography*, Boston, MA, David R. Godine, Publisher, Inc., 1988.

Goudy, Frederick, *Goudy's Type Designs [Complete]: His Story and His Specimens*, New Rochelle, NY, The Myriade Press, 1978.

Grauel, Ralf, *People*, Würzburg, Germany, Koenig & Bauer Group, 2017.

Greer, Carl Richard, *Advertising and its Mechanical Production*, New York, NY, Thomas Y. Crowell Company, 1931.

Hackleman, Charles, *Commercial Engraving and Printing*, Indianapolis, Indiana, Commercial Engraving Publishing Company, 1924.

Hamilton, LL. D., Frederick W., *A Brief History of Printing: Part I: The Development of the Industry, The Great Pioneers*, Chicago, IL, United Typothetae of America, 1918.

Hamilton, LL. D., Frederick W., *A Brief History of Printing: Part II: The Economic History of Printing*, Chicago, IL, United Typothetae of America, 1918.

Hamilton, LL. D., Frederick W., *A Brief History of Printing in America*, Chicago, IL, United Typothetae of America, 1918.

Headrick, Daniel R., *Technology: A World History*, New York, NY, Oxford University Press, 2009.

Hegel, G.W.F., *Science of Logic: Volume Two*, New York, NY, Humanities Press, 1966.

Heller, Steven and Philip B. Meggs, *Texts on Type: Critical Writings on Typography*, New York, NY, Allworth Press, 2001.

Herkimer County Historical Society, *The Story of the Typewriter: 1873-1923*, Herkimer, NY, Herkimer County Historical Society, 1923.

Hiltzik, Michael A., *Dealers of Lightning: Xerox PARC and the Dawn of the Computer Age*, New York, NY, Harper, 1999.

Hind, Arthur M., *A History of Engraving and Etching from the 15th Century to the Year 1914*, London, Constable and Company, Ltd., 1923.

Hindle, Brooke, *The Pursuit of Science in Revolutionary America*, Chapel Hill, NC, The University of North Carolina Press, 1956.

Hindle, Brooke, and Steven Lubar, *Engines of Change: The American Industrial Revolution, 1790–1860*, Washington, D.C., Smithsonian Institution Press, 1986.

Hoe, Robert, *A Short History of the Printing Press and of the Improvements in Printing Machinery from the Time of Gutenberg Up to the Present Day*, New York, NY, Robert Hoe, 1902.

Howard, Robert, *Connecting the Dots: My Life and Inventions, From X-Rays to Death Rays*, New York, NY, Welcome Rain Publishers, 2009.

Hunter, Dard, *Paper-Making in the Classroom*, Peoria, Illinois, The Manual Arts Press, 1931.

Hunter, Dard, *Papermaking: The History and Technique of an Ancient Craft*, New York, NY, Dover Publications, Inc., 1974.

Isaacson, Walter, *Benjamin Franklin: An American Life*, New York, NY, Simon and Schuster, 2003.

Isaacson, Walter, *The Innovators: How a Group of Hackers, Geniuses, and Geeks Created the Digital Revolution*, New York, NY, Simon and Schuster, 2014.

Isaacson, Walter, *Steve Jobs*, New York, NY, Simon and Schuster, 2011.

Itten, Johannes, *The Elements of Color: A Treatise on the Color System of Johannes Itten*, New York, NY, John Wiley & Sons, Inc, 1970.

Ives, Frederic Eugene, *The Autobiography of an Amateur Inventor*, Philadelphia, PA, Press of Innes and Sons, 1928.

Jennett, Seán, *Pioneers in Printing*, London, Routledge and Kegan Paul Limited, 1958.

Johnson, Steven, *How We Got to Now: Six Innovations That Made the Modern World*, New York, NY, Penguin Group, 2014.

Kepler, Michael I.., *Desktop Publishing and Typesetting*, Pittsford, NY, Graphic Dimensions, 1990.

Kipphan, Helmut, *Handbook of Print Media: Technologies and Production Methods*, Heidelberg, Germany, Springer-Verlag, 2001.

Kovarik, Bill, *Revolutions in Communications: Media History from Gutenberg to the Digital Age*, New York, NY, The Continuum International Publishing Group, 2011.

Kuhn, Thomas S., *The Structure of Scientific Revolutions*, Chicago, IL, The University of Chicago Press, 1996.

Kurzweil, Ray, *The Singularity is Near: When Humans Transcend Biology*, New York, NY, Viking Penguin Group, 2005.

Kurzweil, Ray, *The Age of Intelligent Machines*, Cambridge, MA, The MIT Press, 1999.

Langer, Axel, *Story of a Typeface: Helvetica Forever*, Zurich, Switzerland, Lars Müller Publishers GmbH, 2011.

Levenson, Harvey Robert, *Understanding Graphic Communication*, Pittsburgh, PA, GATF Press, 2000.

Levine, I. E., *Miracle Man of Printing: Ottmar Mergenthaler*, New York, NY, Julian Messner, 1966.

Lewis, C. T. Courtney, *George Baxter: His Life and Work*, London, Sampson Low, Marston and Company, Ltd., 1908.

Lilien, Otto M., *History of Industrial Gravure Printing up to 1920*, London, Lund Humphries Publishers Limited, 1972.

Lyons, Martyn, *Books: A Living History*, London, Thames and Hudson, 2011.

McLean, Ruari, *Typographers on Type*, New York, NY, W.W. Norton and Company, 1995.

McLean, Ruari, *Jan Tschichold: Typographer*, Boston, MA, David R. Godine, Publisher, Inc., 1975.

McLean, Ruari, *Manual of Typography*, New York, NY, Thames and Hudson, 1992.

McLuhan, Eric and Frank Zingrone, *Essential McLuhan*, New York, NY, Basic Books, 1995.

McLuhan, Marshall, *The Gutenberg Galaxy: The Making of Typographic Man*, Toronto, ON, University of Toronto Press, 1962.

McLuhan, Marshall, *Understanding Media: The Extensions of Man*, Cambridge, MA, The MIT Press, 1998.

McLuhan, Marshall, *Understanding Me: Lectures and Interviews*, Toronto, ON, The MIT Press, 2005.

McLuhan, Marshall and Bruce R. Powers, *The Global Village: Transformations in World Life and Media in the 21st Century*, New York, NY, Oxford University Press, 1989.

McLuhan, Marshall and Quentin Fiore, *The Medium is the Message: An Inventory of Effects*, Berkley, CA, Ginko Press, 1967.

Mengel, Willi, *Ottmar Mergenthaler and the Printing Revolution*, Brooklyn, NY, Mergenthaler Linotype Company, 1954.

Millionshchikov, M. et al., *The Scientific and Technological Revolution: Social Effects and Prospects*, Moscow, Progress Publishers, 1972.

Milstein, Dalit, *Efi Arazi: His Way, A Biography*, Tel Aviv, Israel, Daniella De-Nur Publishers, 2005.

Minsky, Marvin, *The Society of Mind*, New York, NY, Simon and Schuster, 1986.

Monod, Jacques, *Chance and Necessity*, New York, NY, Vintage Books, 1972.

Montagu, Ashley and Samuel S. Snyder, *Man and the Computer*, Philadelphia, PA, Auerbach Publishers, Inc., 1972.

Morgan, Edmund S., *Benjamin Franklin*, New Haven, CT, Yale University Press, 2003.

Morison, Stanley, *A Tally of Types*, New York, NY, Cambridge University Press, 1973.

Morison, Stanley, *Four Centuries of Fine Printing*, New York, NY, Barnes and Noble, Inc., 1960.

Morison, Stanley, *Letter Forms: Typographic and Scriptorial*, Point Roberts, WA, Hartley and Marks Publishers, Inc., 1997.

Mortimer, Anthony, *Colour Reproduction in a Digital Age*, Surrey, UK, Pira International, 1998.

Müller-Wille, Staffan and Sara Scharf, *Indexing Nature: Carl Linnaeus (1707-1778) and His Fact-Gathering Strategies*, London, London School of Economics, 2009.

Negroponte, *Being Digital*, New York, NY, Alfred A. Knopf, 1995.

Numbers, Ronald L. and Kostas Kampourakis, *Newton's Apple and Other Myths About Science*, Cambridge, MA, Harvard University Press, 2015.

Packer, Randall and Ken Jordan, *Multimedia from Wagner to Virtual Reality*, New York, NY, W. W. Norton and Company, 2002.

Pasko, W. W., *American Dictionary of Printing and Bookmaking*, New York, NY, Burt Franklin, 1970.

Petroski, Henry, *The Evolution of Useful Things*, New York, NY, Vintage Books, 1994.

Pfiffner, Pamela, *Inside the Publishing Revolution: The Adobe Story*, Berkley, CA, The Adobe Press, 2003.

Poe, Marshall T. Poe, *A History of Communications: Media and Society from the Evolution of Speech to the Internet*, New York, NY, Cambridge University Press, 2011.

Poyssick, Gary, *The Politics of Desktop Publishing*, Washington, DC, The Lanman Companies, 1993.

Ralston, Anthony, *Encyclopedia of Computer Science*, New York, NY, Van Nostrand Reinhold Company, 1976.

Roberts, Royston M., *Serendipity: Accidental Discoveries in Science*, New York, NY, John Wiley & Sons, Inc, 1989.

Romano, Frank J., *Machine Writing and Typesetting: The Story of Sholes and Mergenthaler and the Invention of the Typewriter and the Linotype*, Salem, New Hampshire, National Composition Association, 1986.

Romano, Frank J., *Delmar's Dictionary of Digital Printing and Publishing*, Albany, NY, Delmar Publishers, 1997.

Romano, Richard M. and Frank J. Romano, *The GATF Encyclopedia of Graphic Communications*, Sewickley, Pennsylvania, GATF Press, 1998.

Sax, David, *The Revenge of Analog: Real Things and Why They Matter*, New York, NY, Hachette Book Group, 2016.

Schatzberg, Eric, *Technology: Critical History of a Concept*, Chicago, IL, The University of Chicago Press, 2018.

Schlesinger, Carl, *The Biography of Ottmar Mergenthaler: Inventor of the Linotype*, New Castle, DE, Oak Knoll Books, 1989.

Senefelder, Alois, *The Invention of Lithography*, Pittsburgh, PA, GATF Press, 1998.

Sipley, Louis Walton, *A Half Century of Color*, New York, NY, The Macmillan Company, 1951.

Smith, Douglas K. and Robert C. Alexander, *Fumbling the Future: How Xerox Invented, Then Ignored, the First Personal Computer*, Lincoln, NE, toExcel, 1999.

Smith, Larry and Ingrid Swanberg, *D.A. Levy and the Mimeograph Revolution*, Huron, OH, Bottom Dog Press, 2007.

Snyder, Ilana, *Page to Screen: Taking Literacy into the Electronic Age*, New York, NY, Routledge, 1998.

Soderstrom, Walter E., *The Photo-Lithographer's Manual*, New York, NY, Waltwin Publishing Company, 1937.

Sonn, William, *Paradigms Lost: The Life and Deaths of the Printed Word*, Lanham, Maryland, The Scarecrow Press, Inc., 2006.

Sorce, Patricia and Michael Pletka, *Data Driven Print: Strategy and Implementation*, Rochester, NY, RIT Cary Graphic Arts Press, 2006.

Steinberg, S. H., *Five Hundred Years of Printing*, New Castle, DE, The British Library and Oak Knoll Press, 1996.

Suarez, S.J., Michael F. and H.R. Woudhuysen, *The Book: A Global History*, New York, NY, Oxford University Press, 2013.

Tschichold, Jan, *The Form of the Book: Essays on the Morality of Good Design*, Vancouver, BC, Hartley and Marks Publishers, Inc., 1991.

Tschichold, Jan, *The New Typography: A Handbook for Modern Designers*, Los Angeles, CA, University of California Press, 1995.

Usher, Abbott Payson, *A History of Mechanical Inventions*, New York, NY, Dover Publications, Inc., 1988.

Van Wert, I. Gregg and Brett Rutherford, *From Pioneers to Leaders: A History of NAPL's 75 Years of Leadership*, Paramus, NJ, National Association for Printing Leadership, 2007.

Vandendorpe, Christian, *From Papyrus to Hypertext*, Urbana, IL, University of Illinois Press, 1999.

Webb, Joseph W., *Renewing the Printing Industry: Strategies and Action Items for Success*, Harrisville, RI, Strategies for Management, Inc., 2008.

Webb, Joseph W. and Richard M. Romano, *Getting Business: Opportunities for Commercial Printers and Their Clients in the New Communications Arena*, Harrisville, RI, Strategies for Management, Inc., 2011.

Webb, Joseph W. and Richard M. Romano, *Disrupting the Future: Uncommon Wisdom for Navigating Print's Challenging Market Place*, Harrisville, RI, Strategies for Management, Inc., 2010.

Webb, Joseph W. and Richard M. Romano, *This Point Forward: The New Start the Marketplace Demands*, Harrisville, RI, Strategies for Management, Inc., 2014.

Webb, Joseph W. and Richard M. Romano, *The Third Wave*, Wake Forest, NC, Strategies for Management, Inc., 2017.

Weller, Charles E., *The Early History of the Typewriter*, La Porte, IN, Chase E. Shepherd, Printers, 1918.

Wershler-Henry, Darren, *The Iron Whim: A Fragmented History of Typewriting*, Toronto, ON, McLellen and Stewart Ltd., 2005.

Wozniak, Steve, *iWoz: Computer Geek to Cult Icon, How I Invented the Personal Computer, Co-Founded Apple and Had Fun Doing It*, New York, NY, W. W. Norton and Company, 2006.

Wright, Alex, *Cataloging the World: Paul Otlet and the Birth of the Information Age*, New York, NY, Oxford University Press, 2014.

Wroth, Lawrence, *The Colonial Printer*, New York, NY, Dover Publications, Inc., 1965.

Zapf, Hermann, *Future Developments and Alphabet Design*, Rochester, NY, Rochester Institute of Technology, 1969.

Index

About the Author

Kevin Reed Donley has worked in the graphic arts, publishing, and printing industries in the Detroit area for over forty years. He has held positions in all aspects of the business from designer, prepress technician, and customer service representative to sales, marketing, and senior management roles.

Kevin is active in the Detroit and Michigan graphic arts and printing community. He has been a member of the Detroit Club of Printing House Craftsmen and the Advertising Production Club of Detroit and served on the board of directors of the Printing Industries of Michigan and Graphic Media Alliance. Kevin has also been an active participant in the national and international technical standards initiatives of the printing industry. He is a founding member of the GRACoL committee (precursor to the G7 color-matching specification), the pioneering of color reproduction guidelines for commercial offset lithography established in the late 1990s. He is the author of two important industry process management publications—*Roadmaps to Quality in Print: A GRP Travel Guide for Designers and Publishers* (1999) and *BRIDG's Digital Workflow Handbook for Cross Media Publishing* (2002)—to assist the transition from proprietary photomechanical processes to standard digital workflows for content creators and graphic arts service providers during a period of rapid technical change in the printing industry.

Kevin is a speaker on printing business and technology topics and has addressed audiences at colleges and universities and industry events for more than three decades. In 2008, he launched a blog called *Multimediaman: Know the Past, Create the Future* at multimediaman.net and published studies of major innovators in the history of print technology as well as reviews of contemporary developments in new electronic and digital media. These online posts are the source of the chapters in this book as well as the foundation of a theory of technological evolution he has called disruptive continuity.

Kevin was born and raised in Point Pleasant, New Jersey, and graduated from Rutgers University with a degree in Bachelor of Arts in Graphic Design in 1982, after which he relocated to Detroit to start his professional career. In 1997, Kevin became a certified business planning executive with the National Association for Printing Leadership Management Institute at Northwestern University.